NATIONAL NANOTECHNOLOGY INITIATIVE

STRATEGIC PLAN

National Science and Technology Council

Committee on Technology

Subcommittee on Nanoscale Science, Engineering, and Technology

February 2011

National Science and Technology Council
Committee on Technology (CoT)
Subcommittee on Nanoscale Science, Engineering, and Technology (NSET)

CoT Co-Chairs: **Aneesh Chopra**, Office of Science and Technology Policy
Vivek Kundra, Office of Management and Budget
Philip Weiser, National Economic Council

CoT Executive Secretary: **Pedro Espina**, Office of Science and Technology Policy

NSET Subcommittee Co-Chairs:
Travis M. Earles, Office of Science and Technology Policy
Lewis Sloter, Department of Defense

NSET Subcommittee Executive Secretary:
Geoffrey M. Holdridge, National Nanotechnology Coordination Office

National Nanotechnology Coordination Office Director and Coordinator for Standards Development:
E. Clayton Teague

National Nanotechnology Coordination Office Deputy Director and Coordinator for Environment, Health, and Safety Research:
Sally Tinkle

Department and Agency Representatives to the NSET Subcommittee

Office of Science and Technology Policy (OSTP)
Travis M. Earles

Office of Management and Budget (OMB)
Irene B. Kariampuzha

Bureau of Industry and Security (BIS/DOC)
Matt Borman

Consumer Product Safety Commission (CPSC)
Mary Ann Danello
Treye A. Thomas

Department of Defense (DOD)
Akbar Khan
Gernot S. Pomrenke
Lewis Sloter
Eric Snow
David M. Stepp

Department of Education (DOEd)
Krishan Mathur

Department of Energy (DOE)
Harriet Kung
Mihal E. Gross
John C. Miller
Andrew R. Schwartz
Brian G. Valentine

Department of Homeland Security (DHS)
Richard T. Lareau
Eric J. Houser

Department of Justice (DOJ)
Joseph Heaps

Department of Labor (DOL)
Janet Carter

Department of State (DOS)
Ken Hodgkins
Chris Cannizzaro

Department of Transportation (DOT)
Alasdair Cain
Jonathan R. Porter

Department of the Treasury (DOTreas)
John F. Bobalek

Director of National Intelligence (DNI)
Richard Ridgley

Environmental Protection Agency (EPA)
Jeff Morris
Nora F. Savage
Philip G. Sayre

Food and Drug Administration (FDA/DHHS)
Carlos Peña
Ritu Nalubola

Forest Service (FS/USDA)
World L.-S. Nieh
Christopher D. Risbrudt
Theodore H. Wegner

National Aeronautics and Space Administration (NASA)
Michael A. Meador

National Institute of Food and Agriculture (NIFA/USDA)
Hongda Chen

National Institute for Occupational Safety and Health (NIOSH/CDC/DHHS)
Charles L. Geraci
Vladimir V. Murashov

National Institute of Standards and Technology (NIST/DOC)
Lloyd J. Whitman

National Institutes of Health (NIH/DHHS)
Piotr Grodzinski
Lori Henderson
Jeffery A. Schloss

National Science Foundation (NSF)
Mihail C. Roco
Zakya H. Kafafi
Parag R. Chitnis
T. James Rudd

Nuclear Regulatory Commission (NRC)
Stuart Richards

U.S. Geological Survey (USGS)
Sarah Gerould

U.S. International Trade Commission (USITC)
Elizabeth R. Nesbitt

U.S. Patent and Trademark Office (USPTO/DOC)
Charles Eloshway
Bruce Kisliuk

February 4, 2011

Dear Colleagues:

Nanotechnology is generating remarkable scientific and technological advances from the evolutionary to the extraordinary. These advances are enabling a broad spectrum of applications in electronics, medicine, energy, manufacturing, advanced materials, and other fields.

Much of this advancement has been facilitated by the National Nanotechnology Initiative (NNI), which since 2001 has coordinated goals, priorities, and strategies among Federal agencies and promoted groundbreaking interdisciplinary research and infrastructure development critical to nanotechnology innovation. Consistent with the NNI's vision of a future in which nanotechnology benefits society through a revolution in technology and industry, NNI member agencies have plotted a path forward in the 2011 National Nanotechnology Initiative Strategic Plan. Under the auspices of the National Science and Technology Council (NSTC) Subcommittee on Nanoscale Science, Engineering, and Technology (NSET), NNI member agencies will use the plan to guide the coordination of their research, training programs, and resources.

The 2011 NNI strategic plan embodies ten years of U.S. leadership in nanotechnology research and development and illuminates pathways for future breakthroughs. It also benefits from and builds upon recommendations from the President's Council of Advisors on Science and Technology (PCAST) and the National Academies, and incorporates a broad range of stakeholder input obtained through an unprecedentedly open and engaging process. The collective goals and specific objectives articulated in the Strategic Plan will support world-class interdisciplinary nanotechnology research, sustain and expand critical infrastructure, train and inspire the next generation of scientists and engineers, and support responsible development and transfer of nanotechnology into commercial applications to benefit the Nation's economy and the American people.

As the PCAST noted in its 2010 review, the NNI has had a "catalytic and substantial impact" on the growth of nanotechnology innovation in the United States. This plan will help ensure that American leadership in nanotechnology innovation continues into the next decade and beyond.

Sincerely,

JOHN P. HOLDREN
Assistant to the President for Science & Technology
Director, Office of Science and Technology Policy

Table of Contents

Tables and Figures vi

The NNI 1

    Introduction 1

    Vision and Goals 4

    Program Component Areas 5

    NNI Participating Agencies 6

    Relationship Between PCAs and Agency Interests 7

    Agency Interests in Nanotechnology R&D and the NNI 8

Goals and Objectives 23

    Goal 1: Advance a world-class nanotechnology research and development program . . . 23

    Goal 2: Foster the transfer of new technologies into products for commercial and
            public benefit 24

    Goal 3: Develop and sustain educational resources, a skilled workforce, and the
            supporting infrastructure and tools to advance nanotechnology 27

    Goal 4: Support responsible development of nanotechnology. 29

Coordination & Assessment 33

    Nanoscale Science, Engineering, and Technology Subcommittee 33

    Executive Office of the President 37

    Assessment 37

The Path Forward 39

    Collaborative Agency Activities 39

    Anticipated Activities 43

    Developing Partnerships and Engaging Stakeholders 45

    Planned Independent Assessments 45

    Concluding Remarks 45

Appendix A. External Assessment and Stakeholder Input 47

    External Assessment Reports 47

    Stakeholder Input, July–August, 2010 47

    Public Comment on Draft Strategic Plan, November 2010 49

Appendix B. Glossary 50

Tables and Figures

Table 1. Program Component Areas. 5

Table 2. Relationships between the PCAs and the Missions, Interests, and Needs of NNI Agencies . 7

Table 3. Agency Contributions by Thrust Area: Nanotechnology for Solar Energy Signature
Initiative 40

Table 4. Agency Contributions by Thrust Area: Sustainable Nanomanufacturing Signature
Initiative 41

Table 5. Agency Contributions by Thrust Area: Nanoelectronics for 2020 Signature Initiative . . 42

Figure 1. NNI Participating Agencies by Year Joined 6

Figure 2. Coordination and Assessment of the NNI 34

The NNI

The National Nanotechnology Initiative (NNI) is the U.S. Federal Government's interagency program for coordinating research and development and enhancing communication and collaborative activities in nanoscale science, engineering, and technology. This chapter describes the NNI, including the vision and goals that frame the NNI, the categorization of Federal activities in nanotechnology, and the participating NNI agencies.

Introduction

Nanotechnology is the understanding and control of matter at dimensions between approximately 1 and 100 nanometers, where unique phenomena enable novel applications. Work within the intersecting disciplines at the core of nanotechnology innovation—including physical, life, and social sciences and engineering—has revealed the potential of nanomaterials and nanoscale processes to collect and store energy, reinforce materials, sense contaminants, enable life-saving drugs, and shrink and accelerate computational devices in both incremental and paradigm-shifting ways. Further, nanotechnology has enabled development of entirely new materials and devices that can be exploited in each of these and countless other applications.

The United States has set the pace for nanotechnology innovation world-wide with the **National Nanotechnology Initiative (NNI)**. Launched in 2001 with eight agencies, the NNI today consists of the individual and cooperative nanotechnology-related activities of 25 Federal agencies with a range of research and regulatory roles and responsibilities. Fifteen of the participating agencies have research and development (R&D) budgets that relate to nanotechnology, with the reported NNI budget representing the collective sum of these investments. Funding support for nanotechnology R&D stems directly from NNI member agencies, not the NNI. As an interagency effort, the NNI informs and influences the Federal budget and planning processes through its member agencies and through the National Science and Technology Council (NSTC).

Coordinated under the Nanoscale Science, Engineering, and Technology (NSET) Subcommittee of the NSTC's Committee on Technology (CoT), the NNI provides a framework for a comprehensive nanotechnology R&D program by establishing shared goals, priorities, and strategies complementing agency-specific missions and activities and providing avenues for individual agencies to leverage the resources of all participating agencies. Further, the NNI provides a central interface with academia and industry as well as regional/state organizations and international counterparts in the process of innovating nanotechnology. To these ends, the National Nanotechnology Coordination Office (NNCO) provides technical and administrative support to the NSET Subcommittee, serves as a central point of contact for Federal nanotechnology R&D activities, and provides public outreach on behalf of the NNI. Working groups established by the NSET Subcommittee provide an infrastructure to strengthen interagency coordination and collaboration on critical nanotechnology issues.

The ten-year history of U.S. leadership in fundamental nanotechnology research and development under the NNI has established a thriving nanotechnology R&D environment, laid the crucial groundwork for developing commercial applications and scaling up production, and created demand for many new nanotechnology and manufacturing jobs in the near-term. The NNI has dramatically expanded scientific understanding of nanoscale phenomena and enabled the engineering of applications through an extensive and unparalleled infrastructure of R&D centers, networks, and user facilities. The Federal investments in nanotechnology research and development over the past decade have positioned the United States to address key national priorities, bring new expertise to bear on important scientific and social problems, strengthen the social contract between science and society, and inspire a growing number of students to pursue careers in science, technology, engineering, and mathematics. Commercialization resulting from NNI-supported research is mounting.

While the progress of nanotechnology innovations to date has been significant, numerous challenges still exist, and the tremendous potential anticipated from nanoscale R&D is still far from full realization. Exploitation of the full value of nanotechnology innovation depends on sustained fundamental R&D and on focused commercialization efforts. Barriers need to be lowered and pathways streamlined to transfer emerging nanotechnologies into economically viable applications. Researchers, educators, and technicians with new cross-cutting skills are required. Furthermore, at every step there must be a commitment to developing nanotechnology responsibly, with balanced and transparent consideration of the benefits and risks associated with particular nanomaterials in specific applications. For these reasons, broad-based coordination and integration of development efforts across government agencies, academic disciplines, industries, and even countries remain critical to achieving the full economic and societal benefits proven in concept or still promised by nanotechnology.

The **National Nanotechnology Initiative Strategic Plan** is the framework that underpins the nanotechnology work of the NNI member agencies. It aims to ensure that advancements in and applications of nanotechnology R&D to agency missions and the broader national interest continue unabated in this still-young area of research and development. Its purpose is to facilitate achievement of the NNI vision by laying out guidance for agency leaders, program managers, and the research community regarding planning and implementation of nanotechnology R&D investments and activities.

The NSET Subcommittee solicited multiple streams of input to inform the development of this revised NNI Strategic Plan. Independent reviews of the NNI by the President's Council of Advisors on Science and Technology and the National Research Council of the National Academies—strongly supportive of the NNI overall—have made specific recommendations for improving the NNI.[1] Additional input has come from a Strategic Planning Stakeholders Workshop sponsored by the NSET Subcommittee during July 13–14, 2010, as well as from detailed responses to a Request for Information published via the *Federal Register* and from online dialogue in the NNI Strategy Portal.[2] The portal was also used in November 2010, to host a 30 day public comment period on the draft of this plan.

Thus informed by feedback and recommendations from a broad array of stakeholders, this strategic plan represents the consensus of the participating agencies as to the high-level goals and priorities of the NNI and specific objectives for at least the next three years. The strategic plan provides the framework within which each agency will carry out its own mission-related nanotechnology programs and that will sustain coordination of interagency activities. It describes the four overarching goals of the NNI, the

---

[1] See Appendix A for details on external reviews and assessments of the NNI.

[2] Details on the workshop, *Federal Register* notice, and NNI Strategy Portal are available in Appendix A.

major program component areas established in 2004 to broadly track the categories of investments needed to ensure the success of the initiative, and the near-term objectives that will be the concrete steps taken toward collectively achieving the NNI vision and goals. Finally, the plan describes collaborative interagency activities, including nanotechnology signature initiatives that are a new model of specifically targeted and closely coordinated interagency, cross-sector collaboration designed to accelerate innovation in areas of national priority. The first three nanotechnology signature initiatives are focused on renewable energy, sustainable manufacturing, and next-generation electronics.

The 21st Century Nanotechnology Research and Development Act of 2003 calls for the NNI Strategic Plan to be updated triennially; the plan presented here updates and replaces the December 2007 plan.

What is Nanotechnology? *

Nanotechnology is the understanding and control of matter at dimensions between approximately 1 and 100 nanometers, where unique phenomena enable novel applications. Encompassing nanoscale science, engineering, and technology, nanotechnology involves imaging, measuring, modeling, and manipulating matter at this length scale.

A nanometer is one-billionth of a meter. A sheet of paper is about 100,000 nanometers thick; a single gold atom is about a third of a nanometer in diameter. Dimensions between approximately 1 and 100 nanometers are known as the nanoscale. Unusual physical, chemical, and biological properties can emerge in materials at the nanoscale. These properties may differ in important ways from the properties of bulk materials and single atoms or molecules.

The scope of this definition, established by the NNI at its inception, is for identifying and coordinating Federal nanotechnology research and development as well as facilitating communication.

Vision and Goals

The vision of the NNI is *a future in which the ability to understand and control matter at the nanoscale leads to a revolution in technology and industry that benefits society*. The NNI expedites the discovery, development, and deployment of nanoscale science, engineering, and technology to serve the public good, through a program of coordinated research and development aligned with the missions of the participating agencies. In order to realize the NNI vision, the participating agencies are working collectively toward the following four goals:

Goal 1: Advance a world-class nanotechnology research and development program.

The NNI ensures U.S. leadership in nanotechnology research and development by stimulating discovery and innovation. This program expands the boundaries of knowledge and develops technologies through a comprehensive program of research and development. The NNI agencies invest at the frontiers and intersections of many disciplines, including biology, chemistry, engineering, materials science, and physics. The interest in nanotechnology arises from its potential to significantly impact numerous fields, including aerospace, agriculture, energy, the environment, healthcare, information technology, homeland security, national defense, and transportation systems.

Goal 2: Foster the transfer of new technologies into products for commercial and public benefit.

Nanotechnology contributes to U.S. competitiveness and national security by improving existing products and processes and by creating new ones. The NNI implements strategies that maximize the economic benefits of its investments in nanotechnology, based on understanding the fundamental science and responsibly translating this knowledge into practical applications.

Goal 3: Develop and sustain educational resources, a skilled workforce, and the supporting infrastructure and tools to advance nanotechnology.

A skilled science and engineering workforce, leading-edge instrumentation, and state-of-the-art facilities are essential to advancing nanotechnology research and development. Educational programs and resources are required to produce the next generation of nanotechnologists—that is, the researchers, inventors, engineers, and technicians who drive discovery, innovation, industry, and manufacturing.

Goal 4: Support responsible development of nanotechnology.

The NNI aims to maximize the benefits of nanotechnology and at the same time to develop an understanding of potential risks and to develop the means to manage them. Specifically, the NNI pursues a program of research, education, collaboration, and communication focused on environmental, health, safety, and broader societal dimensions of nanotechnology development. Responsible development requires engagement with universities, industry, government agencies (local, state, and Federal), nongovernmental organizations, and other communities.

Program Component Areas

Program component areas (PCAs) are major subject areas under which are grouped related nanotechnology R&D projects and activities. They provide an organizational framework for categorizing the activities of the NNI. Investment and progress in these areas is critical to achieving the NNI's goals and to realizing its vision. The investment related to each PCA is reported in the annual NNI supplement to the President's Budget.[3] The eight PCAs are described in Table 1. Agency projects and activities in one or more PCAs are critical to progress toward realizing each goal.

Table 1. Program Component Areas

No.	PCA Title	Description
1	Fundamental Nanoscale Phenomena and Processes	Discovery and development of fundamental knowledge pertaining to new phenomena in the physical, biological, and engineering sciences that occur at the nanoscale. Elucidation of scientific and engineering principles related to nanoscale structures, processes, and mechanisms.
2	Nanomaterials	Research aimed at the discovery of novel nanoscale and nanostructured materials and at a comprehensive understanding of the properties of nanomaterials (ranging across length scales, and including interface interactions). R&D leading to the ability to design and synthesize, in a controlled manner, nanostructured materials with targeted properties.
3	Nanoscale Devices and Systems	R&D that applies the principles of nanoscale science and engineering to create novel, or to improve existing, devices and systems. Includes the incorporation of nanoscale or nanostructured materials to achieve improved performance or new functionality. To meet this definition, the enabling science and technology must be at the nanoscale, but the systems and devices themselves are not restricted to that size.
4	Instrumentation Research, Metrology, and Standards for Nanotechnology	R&D pertaining to the tools needed to advance nanotechnology research and commercialization, including next-generation instrumentation for characterization, measurement, synthesis, and design of materials, structures, devices, and systems. Also includes R&D and other activities related to development of standards, including standards for nomenclature, materials characterization and testing, and manufacture.
5	Nanomanufacturing	R&D aimed at enabling scaled-up, reliable, and cost-effective manufacturing of nanoscale materials, structures, devices, and systems. Includes R&D and integration of ultra-miniaturized top-down processes and increasingly complex bottom-up or self-assembly processes.
6	Major Research Facilities and Instrumentation Acquisition	Establishment of user facilities, acquisition of major instrumentation, and other activities that develop, support, or enhance the nation's scientific infrastructure for the conduct of nanoscale science, engineering, and technology R&D. Includes ongoing operation of user facilities and networks.
7	Environment, Health, and Safety	Research primarily directed at understanding the environmental, health, and safety impacts of nanotechnology development and corresponding risk assessment, risk management, and methods for risk mitigation.
8	Education and Societal Dimensions	Education-related activities such as development of materials for schools, undergraduate programs, technical training, and public communication, including outreach and engagement. Research directed at identifying and quantifying the broad implications of nanotechnology for society, including social, economic, workforce, educational, ethical, and legal implications.

---

[3] All annual supplements are available at http://nano.gov.

NNI Participating Agencies

Federal Agencies with Budgets Dedicated to Nanotechnology Research and Development

Consumer Product Safety Commission (CPSC)
Department of Defense (DOD)
Department of Energy (DOE)
Department of Homeland Security (DHS)
Department of Justice (DOJ)
Department of Transportation (DOT, including the Federal Highway Administration, FHWA)
Environmental Protection Agency (EPA)
Food and Drug Administration (FDA, Department of Health and Human Services)
Forest Service (FS, U.S. Department of Agriculture)
National Aeronautics and Space Administration (NASA)
National Institute for Occupational Safety and Health (NIOSH, Department of Health and Human Services)
National Institute of Food and Agriculture (NIFA, U.S. Department of Agriculture)[4]
National Institutes of Health (NIH, Department of Health and Human Services)
National Institute of Standards and Technology (NIST, Department of Commerce)
National Science Foundation (NSF)

Other Participating Agencies

Bureau of Industry and Security (BIS, Department of Commerce)
Department of Education (DOEd)
Department of Labor (DOL, including the Occupational Safety and Health Administration, OSHA)
Department of State (DOS)
Department of the Treasury (DOTreas)
Intelligence Community (IC)
Nuclear Regulatory Commission (NRC)
U.S. Geological Survey (USGS, Department of the Interior)
U.S. International Trade Commission (USITC)[5]
U.S. Patent and Trademark Office (USPTO, Department of Commerce)

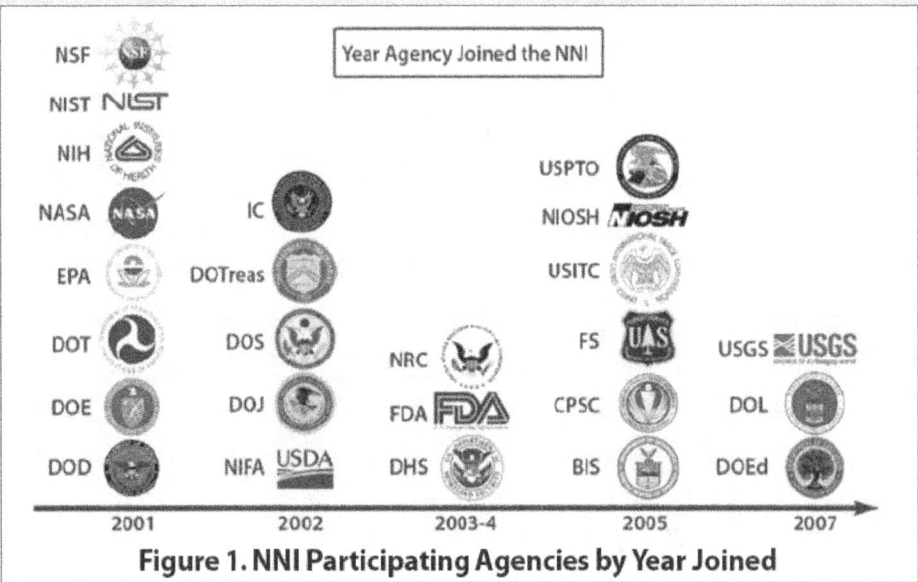

Figure 1. NNI Participating Agencies by Year Joined

---

[4] Formerly the Cooperative State Research, Education, and Extension Services (CSREES).
[5] Observer status.

Relationship Between PCAs and Agency Interests

The NNI program component areas (PCAs) cut across the interests and activities of the participating agencies and represent areas where achieving the goals of the NNI can be expedited through inter-agency coordination. Table 2 shows, for each participating agency, which PCAs have the strongest relationships to the agency's mission, interests, and needs. The strength of the relationships shown may correlate with the level of that agency's investment. However, in some cases—especially for those agencies that do not have nanotechnology R&D budgets—there are nevertheless strong connections between PCAs and agency missions.

Table 2. Relationships between the PCAs and the Missions, Interests, and Needs of NNI Agencies

	Fundamental Nanoscale Phenomena & Processes	Nanomaterials	Nanoscale Devices & Systems	Instrumentation Research, Metrology, & Standards	Nanomanufacturing	Major Research Facilities & Instrumentation Acquisition	Environment, Health, & Safety	Education & Societal Dimensions
BIS (DOC)	•	O	O	O	•			
CPSC	•	•	O	O	•		O	•
DOD	O	O	O	•	O	•	•	•
DOEd							•	O
DOE	O	O	•	•	•	O	•	•
DHS	•	•	O	O	•	•	•	
DOJ/NIJ			O					•
DOL		•			•		O	O
DOS	•	•	•	•	•	•	O	O
DOT	O	O	O		•		•	
DOTreas		O	O					
EPA	•	O	O	•	O		O	•
FDA (DHHS)	•	•	•	•	•		O	
FS (USDA)	•	O	O	•	O		•	
IC/DNI	O	O	O		O			
NASA	•	O	O		•	•		
NIFA (USDA)	O	O	O	•	•		O	O
NIH (DHHS)	O	O	O	•	•	•	O	•
NIOSH (DHHS)		•			•		O	•
NIST (DOC)	O	O	•	O	O	O	•	•
NSF	O	O	O	•	O	O	O	O
NRC		O	•					
USGS (DOI)	O				O		O	
USITC		O	O		O			
USPTO (DOC)		O	O	O	O			O

O Primary   • Secondary

Agency Interests in Nanotechnology R&D and the NNI

In August 2000, the Subcommittee on Nanoscale Science, Engineering, and Technology (NSET) was constituted as part of the NSTC CoT specifically to facilitate interagency collaboration on nanoscale R&D and to provide a framework for setting Federal R&D budget priorities related to nanotechnology. The NSET Subcommittee member agencies continue to fund nanoscale science and engineering research because the work done so far supports the early assumptions about the value of this growing scientific endeavor. Moreover, the platform for communication, collaboration, and coordination provided by the NNI through the NSET Subcommittee continues to foster the engagement of all member agencies, including those with an interest, though no targeted funding, in nanotechnology. The agencies describe below their individual interests in nanotechnology R&D and the value of the NNI, as they collectively contribute by various means to the welfare of the nation and to their respective agency missions and responsibilities.[6]

Bureau of Industry and Security, Department of Commerce

The interagency coordination provided by the NNI enables the Bureau of Industry and Security of the Department of Commerce (BIS/DOC) to stay apprised of new nanotechnology advancements that may present national security challenges and that may provide opportunities for companies in the national defense industrial base. Further, the NNI creates mechanisms (i.e., through regular meetings of the NSET Subcommittee) for BIS to share information about national security needs and challenges with other Federal agencies. BIS may also exercise its statutory data collection authority, as needed in support of the NNI vision. Together, these exchanges support the BIS mission to advance U.S. national security, foreign policy, and economic objectives by ensuring an effective export control and treaty compliance system and promoting continued U.S. strategic technology leadership.

Consumer Product Safety Commission

The Consumer Product Safety Commission (CPSC) in cooperation with Federal partners analyzes the use and safety of nanotechnology in consumer products. In order to meet identified data needs, the CPSC staff has met with and collaborates with staff at a number of Federal agencies in areas of mutual interest where collaboration would be beneficial and support the respective missions of each agency. More consumer products are using compounds or materials that have been produced using nanotechnologies that directly manipulate matter at the atomic level and fabrication of materials that could not have been produced in the past. Although these nanomaterials may have the same chemical composition as non-nanomaterials, at the nanoscale these nanomaterials may demonstrate different physical and chemical properties and they may behave differently in the environment and the human body. CPSC developed an internal nanotechnology team comprised of various technical experts (e.g., engineers, toxicologists, and economists) to advise the Commission on the safe use of nanotechnology in consumer products. As part of the NNI, the CPSC nanotechnology team participates in the interagency collection and analysis of data and the development of reports that focus upon the potential environmental, health, and safety issues associated with the use of nanotechnology.

---

[6] The most updated links to nanotechnology activities at NNI participating agencies can be found at http://nano.gov.

Department of Defense

Department of Defense (DOD) leadership considers nanotechnology to have high and growing potential to contribute to the warfighting capabilities of the nation. Because of the broad and interdisciplinary nature of nanotechnology, DOD leadership views it as an enabling technology area that should receive the highest level of department attention and coordination. The vision and capability construct of Defense Research and Engineering includes nanotechnology as one of four exemplary foundational technologies, along with advanced materials, advanced electronics, and manufacturing technology. DOD Basic Research acknowledges that realizing the potential of nanotechnology is a key research objective. In particular, nanotechnology is an enabling technology for new classes of sensors (such as novel focal plane arrays and chemical/biological threat sensors), communications, and information processing systems needed for qualitative improvements in persistent surveillance. The DOD also invests in nanotechnology for advanced energetic materials, photocatalytic coatings, active microelectronic devices, structural fibers, strength- and toughness-enhancing additives, advanced processing, and a wide array of other promising applications. The DOD nanotechnology efforts are based on coordinated planning and federated execution among the military departments and agencies (e.g., the Defense Advanced Research Projects Agency and the Defense Threat Reduction Agency). Although DOD does not establish funding targets for nanotechnology specifically, its support for nanotechnology-related research and development has continued to increase through its competitive success in core research planning, technology development solicitations, and Federal programs such as Small Business Innovation Research (SBIR) and the Multidisciplinary University Research Initiative.

DOD was among the initial participating agencies in the NNI and the NSET Subcommittee. The department considers the initiative and its formal coordination fora to have been and to continue to be valuable as a means to facilitate technology planning, coordination, and communication among the Federal agencies. The meetings and workshops hosted or facilitated by the NNI participants help to identify and define options and opportunities that materially contribute to DOD planning activities and program formulation. The reviews and collegial meetings, working groups, and task forces established under the auspices of the NSET Subcommittee are valuable means of formal and informal coordination at the Federal level and form a solid basis for exploring collaborative activities, addressing mutual or pervasive issues, and identifying areas in which interagency assistance is needed or would be productive. The DOD has continuously contributed to the NNI through participation in the above-noted activities and through numerous outreach and programmatic efforts in which nanotechnology has been a principal aspect of the program or planning. The transparency that is enabled by the NNI is viewed as symmetrically beneficial to DOD, the other agencies, and the many private-sector stakeholders in the broad arena of nanoscience, nanotechnology, and nanotechnology-enabled applications.

Department of Education

The Department of Education (DOEd) faces major challenges in a number of education-related areas, including a need for more graduates and researchers in areas of science, technology, engineering, and mathematics (STEM) education. By providing working groups, regular NSET Subcommittee meetings, and inter-agency communication channels, the NNI provides a mechanism for DOEd to better collaborate with other relevant agencies, such as the National Science Foundation (NSF), which makes substantial investments in nanotechnology-related education, and the Department of Labor (DOL), which follows trends in workforce needs.

Department of Energy

Department of Energy (DOE) leadership views nanoscience and nanotechnology as having a vitally important role to play in solving the energy and climate-change challenges faced by the nation. This broad and diverse field of research and development will likely have dramatic impact on future technologies for solar energy collection and conversion, energy storage, alternative fuels, and energy efficiency, to name just a few. DOE has participated in the NNI since its inception and maintains a strong commitment to the initiative, which has served as an effective and valuable way of spotlighting needs and targeting resources in this critical emerging area of science and technology. The NNI continues to provide a focus for overall investment in physical sciences, a crucial locus for interagency communication and collaboration, and an impetus for coordinated planning. The research and infrastructure successes spurred by the NNI have made the United States a world leader in this area, with significant national benefit.

DOE funding spans all eight program component areas of the NNI, with the majority falling into three categories: fundamental phenomena and processes (PCA 1), nanomaterials (PCA 2), and major research facilities and instrumentation acquisition (PCA 6). In the latter category, the DOE investment is significantly larger than that of any other agency, due primarily to the planning, construction, and operation of Nanoscale Science Research Centers (NSRCs) located at DOE laboratories. The NSRCs operate as user facilities, with access based on submission of proposals that are reviewed by independent evaluation boards, and at no cost for nonproprietary work. The NSRCs support synthesis, processing, fabrication, and analysis at the nanoscale and are designed to be state-of-the-art user centers for interdisciplinary nanoscale research, serving as an integral part of DOE's comprehensive nanoscience program that encompasses new science, new tools, and new computing capabilities.

Department of Homeland Security

Department of Homeland Security (DHS) interests in nanoscience are primarily focused on the application of nanoscale materials and devices that provide enhancements in component technology performance for homeland security applications. The applications for the efforts described below are in threat detection for enhanced security for aviation, mass transit, and first responders:

- *Materials toolbox:* These efforts are focused on the development of materials systems that allow systematic control of chemical and structural features from molecular scales (functional groups) through nano- and microscales. The ability to precisely tune material properties is critical for successful development of improved active sensor surfaces and analyte collection substrates as well as development of novel sensing structures and arrays.

- *Advanced preconcentrators:* The DHS Science and Technology Directorate is currently investigating the development of high-performance preconcentrators for use in next-generation detection systems. The focus of these efforts is the development of nano- and microscale materials that enable radio-frequency and optical control of device temperature. To date, several functional prototypes have been demonstrated. Commercialization of these devices is currently being pursued.

- *Advanced sensing platforms:* Work on the development of multimodal carbon nanotube sensing platforms continues with industry partners. The emphasis of these efforts is on development of manufacturing techniques for low-cost sensor platforms.

Department of Justice/National Institute of Justice

The National Institute of Justice (NIJ) investment in nanotechnology furthers the Department of Justice's (DOJ) mission through the sponsorship of research that provides objective, independent, evidence-based knowledge and tools to meet the challenges of crime and justice, particularly at the state and local levels. New projects are awarded on a competitive basis, and therefore, total investment may change each fiscal year. However, NIJ continues to view nanotechnology as an integral component of its research and development portfolio as applicable to criminal justice needs.

Department of Labor/ Occupational Safety and Health Administration

The Department of Labor (DOL) Occupational Safety and Health Administration (OSHA) plays an integral role in nanotechnology by protecting the nation's workforce. Through the NNI interagency efforts, OSHA accomplishes its mission by collaborating and sharing information with other Federal agencies. As part of this effort, OSHA's goal is to educate employers on their responsibility to protect workers and educate them on safe practices in handling nanomaterials. OSHA is developing guidance and educational materials promoting worker safety and health that will be shared with the public and through the NNI.

In addition, OSHA is interested in ensuring responsible and sustainable nanotechnology by promoting and developing manufacturing processes that take safety and health into consideration from the design of manufacturing systems throughout the entire lifecycle of the material in use, wherever there is potential for worker exposure. To achieve this objective, OSHA is participating in the nanomanufacturing signature initiative and collaborating with Environmental Protection Agency (EPA) and the National Institute for Occupational Safety and Health (NIOSH) to promote sustainable development in the manufacturing process. This involves development of nanomanufacturing processes that take into account exposure control measures to eliminate or reduce worker exposure from the outset.

Department of State

The Department of State (DOS) actively participates in the NNI in order to identify and promote multilateral and bilateral scientific activities that support U.S. foreign policy objectives, protect national security interests, advance economic interests, and foster environmental protection. International scientific collaboration enhances existing U.S. research, development, and innovation programs. Nanotechnology's enormous potential to address global challenges relating to water, health, and energy renders it an ideal subject for collaboration on pre-competitive and non-competitive research. DOS assists NNI member agencies to establish partnerships with counterpart institutions abroad by holding regular joint committee meetings with representatives from over forty countries. These meetings are governed by binding science and technology agreements that facilitate exchange of scientific results, provide for protection and allocation of intellectual property rights and benefit sharing, facilitate access for researchers, address taxation issues, and respond to the complex set of issues associated with economic development, domestic security, and regional stability. More broadly, through chairmanship of the NSET Subcommittee's Global Issues in Nanotechnology (GIN) Working Group, DOS coordinates U.S. Federal

Government interactions with foreign governments and multilateral institutions to foster mutually beneficial cooperation on nanoscale science and technology, to develop an international marketplace for nanotechnology products and ideas, and to establish a framework for the safe, secure, and responsible use of nanotechnology. DOS also leads efforts in the Working Party on Nanotechnology (WPN) of the Organisation for Economic Co-operation and Development (OECD), the Strategic Approach to International Chemicals Management (SAICM), and other international organizations to communicate these precepts globally to key policymakers and stakeholders.

Department of Transportation

The Department of Transportation's (DOT) Federal Highway Administration (FHWA) sees great promise in the application of nanotechnology to help solve long-term highway and transportation research needs in support of DOT's strategic goals: Safety, Livable Communities, State of Good Repair, Economic Competitiveness, and Environmental Sustainability. By strategically investing in focused research areas and leveraging investments in nanoscale technology by other NNI partners and Federal agencies, industry, and academia, FHWA aims to accelerate the capability to provide safer, more efficient, longer-lasting highway transportation systems. Based on the findings of a March 2009 workshop of experts from academia, DOT, and other Federal agencies, FHWA's Exploratory Advanced Research Program is investing in nanoscale research to address key highway research issues in infrastructure, safety, operations, and the environment. Nanotechnology promises breakthroughs in multiple areas, offering a potential for synergy and benefits across many traditional highway research focus areas.

The development of innovative materials and coatings can deliver significant improvements in durability, performance, and resiliency of highway and transportation infrastructure components. Nanoscale engineering of traditional transportation infrastructure materials (e.g., steel, concrete, asphalt, and other cementitious materials, as well as recycled forms of these materials) offers great promise. Developments in nanoscale sensors and devices may provide cost-effective opportunities to embed and employ structural health monitoring systems to continuously monitor corrosion, material degradation, and performance of structures and pavements under service loads and conditions. In addition, these developments might provide multifunctional properties to traditional infrastructure materials, such as the ability to generate or transmit energy. Nanoscale sensors and devices may also enable a cost-effective infrastructure that communicates with vehicle-based systems to assist drivers with tasks such as maintaining lane position, avoiding collisions at intersections, and modifying or coordinating travel behavior to mitigate congestion or adverse environmental impacts. Other environmental applications include sensors to monitor mobile source pollutants and air, water, and soil quality.

FHWA's long-term strategy is to continue targeted investment in select areas while building an appreciation for highway research needs with NNI partners and the broader nanoscale research community in order to augment longstanding partnerships and make significant progress toward improving the nation's highway and transportation systems.

Department of Treasury

The Department of the Treasury (DOTreas) works through the NSET Subcommittee to help the NNI achieve its vision congruent with the DOTreas: to serve the American people and strengthen national security by managing the U.S. Federal Government's finances effectively, to promote economic growth and stability, and to ensure the safety, soundness, and security of U.S. and international financial systems. DOTreas monitors those aspects of developing nanotechnology that could most effectively assist the execution of its role as the steward of the U.S. economic and financial systems, and as an influential participant in the global economy. DOTreas seeks to assess and utilize nanotechnology in the discharge of its responsibilities, including advising the President on economic and financial issues, encouraging sustainable economic growth, and fostering improved governance in financial institutions. It seeks to harness those aspects of nanotechnology R&D that will allow it to better operate and maintain systems that are critical to the nation's financial infrastructure, such as the production of coin and currency. Interactions with the NSET Subcommittee help DOTreas as it endeavors to capture developments in nanoscale science and engineering that are changing the parameters of its domestic and international operations, particularly those impacting its critical national security-related activities in implementing economic sanctions against foreign threats to the United States, identifying and targeting the financial support networks of national security threats, improving the safeguards of U.S. financial systems, and creating new economic and job opportunities to promote economic growth and stability at home and abroad.

Environmental Protection Agency

The Environmental Protection Agency (EPA) has a dual interest in nanotechnology for the protection of human health and the environment. First, EPA is interested in understanding the potential implications of engineered nanomaterials, including understanding how nanomaterials can be designed and used in ways that minimize any adverse public health or environmental impacts. Second, the EPA is interested in the potential of nanotechnology to improve the environment, including its use for environmental sensing, remediating pollutants, and for replacing more-toxic substances. Both interests have foundations in the theme of achieving sustainability in use of nanotechnology.

Potentially, nanotechnology offers transformational capabilities for a vast array of products and processes, including those that enhance environmental quality and sustainability. To help nanotechnology create maximum societal benefits and to minimize its potential environmental impacts, EPA works with its Federal partners within the NSET Subcommittee to ensure that research gaps are covered, critical issues are addressed, and information is communicated to all interested stakeholders.

Food and Drug Administration

Nanomaterials often have chemical, physical, or biological properties that are different from those of conventional materials. Such differences may include altered magnetic, electrical, or optical properties, structural integrity, and chemical or biological activity. Because of researchers' ability to engineer such properties, nanomaterials have great potential for use in a vast array of products, including products regulated by the Food and Drug Administration (FDA). Also, because of some of their special properties, nanomaterials may pose different or additional issues for toxicologic, safety, and effectiveness assessments. As such, there is a growing need for scientific information and tools to help better predict or detect the potential impact of nanomaterials on human and animal health.

FDA nanotechnology investments are focused on enabling the agency to characterize nanotechnology-based products, develop models for safety and effectiveness assessment, and study the behavior of nanomaterials in biological systems and their effects on human health. These investments support

FDA's mission to protect and promote public health and help ensure the responsible development of nanotechnology.

FDA also continues to foster and develop collaborative relationships with other Federal agencies through participation in the NNI and the NSET Subcommittee, as well as with sister regulatory agencies, international organizations, healthcare professionals, industry, consumers, and other stakeholders. These collaborations allow information to be exchanged efficiently and serve to identify research needs related to the use of nanomaterials in FDA-regulated products. Although FDA activities are relevant to all four NNI goals, FDA efforts are primarily focused on Goal 4, to facilitate responsible development of nanotechnology, in three FDA priority areas: (1) building laboratory and product testing capacity, (2) establishing scientific staff development and training, and (3) engaging in collaborative and interdisciplinary research to address product characterization and safety.

Forest Service, U.S. Department of Agriculture

Nanotechnology has enormous promise to bring about fundamental changes in and significant benefit from our nation's use of renewable resources. For example, cellulose nanofibers and cellulose nanocrystals derived from trees: (1) are renewable; (2) are produced in trees via photosynthesis from solar energy, atmospheric carbon dioxide, and water; (3) store carbon; and (4) depending upon how long cellulose-based products remain in service, are carbon negative or carbon neutral. These cellulosic nanomaterials have strength properties greater than Kevlar®, have piezoelectric properties equivalent to quartz, and can be manipulated to produce photonic structures. The USDA Forest Service's (FS/USDA) Forest Products Laboratory, in collaborations with Purdue University and others, has been conducting research on characterization, predictive modeling, surface modification, and sensor applications of cellulose nanocrystals. Current global research directions in cellulose nanomaterials indicate that this material could be used for a variety of new and improved product applications such as lighter and stronger paper and paperboard products; lighter and stronger building materials; wood products with improved durability; barrier coatings; body armor; automobile and airplane composite panels; electronics; biomedical applications; and replacement of petrochemicals in plastics and composites. The U.S. forest products industry, the major supplier and a user of cellulose nanomaterials, through the American Forest & Paper Association Agenda 2020 Technology Alliance, has initiated a Cooperative Board for Advancing Nanotechnology (CBAN).

Through participation in the NNI and representation on the NSET Subcommittee, USDA Forest Service R&D has begun partnering with other Federal entities (e.g., NIST, NSF, DOE, DOD), industry, and academia to develop the pre-competitive science and technology critical to the economic and sustainable production and use of new high-value, nanotechnology-enabled forest-based products. Participation in the NNI and the NSET Subcommittee has helped create a favorable environment for increased Forest Service investment in nanotechnology R&D. Forest Service nanotechnology research has contributed broadly to the NNI program component areas, with primary emphasis on fundamental nanoscale phenomena and processes (PCA 1); nanomaterials (PCA 2); nanoscale devices and systems (PCA 3); instrumentation research, metrology, and standards (PCA4); and nanomanufacturing (PCA5); with possible future investment in environment, health, and safety (PCA 7).

Intelligence Community/Director of National Intelligence

There are several agencies within the intelligence community (IC) that conduct nanotechnology research and development. The National Reconnaissance Office (NRO) has an R&D program that focuses on nanoelectronics, nanomaterials, and energy generation and storage using nanotechnologies.

In nanoelectronics, both analog and digital, the emphasis is on ultralow power for terrestrial data centers and radiation-hardened ultralow power for satellites. Carbon-based nanoelectronics are compatible with today's microelectronics and the foundries that produce them. A major focus going forward will be on ultradense, ultralow-power nonvolatile memory for saving power in data centers and satellites, replacement for today's silicon logic, and advanced linear analog nanoelectronics for next-generation communications and radar systems. These nanoelectronics will transform today's systems into advanced capabilities that will solve tomorrow's IC challenges.

Nanomaterials, both carbon-based sheets and threads, will be used to develop advanced ultralight, ultrastrong composites for satellites, unmanned aircraft, and advanced body armor. Carbon-based threads will be used to develop novel ultralightweight cables and wires for satellites, aircraft, and data centers that reduce weight by as much as 80% and deliver more data signals and power than conventional copper wires and cables.

Nanotechnologies are being applied to solar cells to achieve 35% near-term efficiency and develop 40% to 47% efficiencies in the mid-term for use in space. With the application of 10 to 1000 times normal sunlight (concentration), 52% to 61% efficiency can be achieved for terrestrial use, as defined by current research. Carbon-based nanomaterials are also being developed for advanced lithium ion batteries with 3–5 times more power, more rapid rechargeability, and much lighter weight than current lithium ion batteries.

Nanotechnology provides the IC with transformative and game-changing capabilities not achievable with conventional electronics, materials, or power technologies, and with greatly reduced size, weight, and power. The NSET Subcommittee provides an open forum where agencies can describe their nanotechnology portfolios to other agencies, making them aware of progress achieved. It also affords the opportunity to collaborate to further accelerate nanotechnology R&D, prototyping, nanomanufacturing, *in situ* and post-product metrology, and final transition to acquisition programs.

National Aeronautics and Space Administration

The three prime drivers for the National Aeronautics and Space Administration's (NASA) aerospace research and development activities are to (1) reduce vehicle weight, (2) enhance performance, and (3) improve safety, durability, and reliability. Nanotechnology is a tool to address each of these drivers. Nanotechnology research at NASA is focused in four areas: engineered materials and structures; energy generation, storage, and distribution; electronics, sensors, and devices; and propulsion. This research is conducted through a combination of in-house activities at NASA research and flight centers, competitively funded research with universities and industry, and collaborations with other agencies, universities, and industry. Through the University Research Centers Program, NASA has also funded nanotechnology research at minority-serving institutions, including the Center for Advanced Nanoscale

Materials at the University of Puerto Rico and the High Performance Polymers and Composites Center at Clark Atlanta University.

NASA has participated in the NNI since its inception and is committed to partnering with other member agencies to identify key technical challenges in nanotechnology R&D, focus resources to address these challenges, and accelerate the development of nanotechnology breakthroughs and their translation into commercial products.

National Institute of Food and Agriculture, U.S. Department of Agriculture

The National Institute of Food and Agriculture (NIFA) of the U.S. Department of Agriculture (USDA), established by the 2008 Farm Bill, serves the nation's needs by supporting exemplary research, education, and extension to address challenges. NIFA's mission is to lead food and agricultural sciences to help create a better future for the nation and the world. NIFA's current priority areas are (1) global food security and hunger, (2) climate change, (3) sustainable bioenergy, (4) nutrition and childhood obesity, and (5) food safety. Nanoscale science, engineering, and technology have demonstrated their relevance and great potential to enable revolutionary improvements in agriculture and food systems, including plant production and products; animal health, production, and products; food safety and quality; nutrition, health, and wellness; renewable bioenergy and biobased products; natural resources and the environment; agriculture systems and technology; and agricultural economics and rural communities.

NIFA's predecessor agency was among the early participating agencies in the NSET Subcommittee[7], joining in 2002, and that agency (later, NIFA) has actively participated in and contributed to NNI activities ever since. The NNI provides a solid platform on which NIFA can effectively explore opportunties in nanoscience and nanotechnology to address critical societal challenges facing agriculture and food systems through coordination, collaboration, and leveraging resources with other Federal agencies. Scientific discoveries and technological breakthroughs inspire agricultural and food scientists to seek novel solutions. The extensive infrastructure networks developed by the NNI enhance the productivity and expand the capability of agricultural and food science research and development in academia and industry. NIFA actively contributes to and benefits significantly from its participation in the NNI activities to identify research gaps and opportunities through workshops and discussions, to support public engagement and communication, to facilitate public-private partnerships in close collaboration with industry, and to participate in and promote international information exchanges and cooperation. NIFA also supports multiagency joint research efforts of common interest and importance as appropriate to its mission, goals, and objectives. The agency's nanotechnology programs have broadly contributed to the NNI, with primary emphasis on fundamental nanoscale phenomena and processes (PCA 1); nanomaterials (PCA 2); nanoscale devices and systems (PCA 3); environment, health, and safety (PCA 7); and education and societal dimensions (PCA 8). NIFA's SBIR program also supports innovative nanotechnology R&D throughout its broad topic areas.

National Institutes of Health

The National Institutes of Health (NIH), a part of the U.S. Department of Health and Human Services (DHHS), is the primary Federal agency for conducting and supporting medical research. The NIH mission

---

[7.] NIFA participated in the NNI as the USDA's Cooperative State Research, Education, and Extension Service (CSREES) until it was reorganized and renamed in 2009.

is to seek fundamental knowledge about the nature and behavior of living systems and the application of that knowledge to enhance health, lengthen life, and reduce the burdens of illness and disability. Toward this end, NIH leadership realizes that advances in nanoscience and nanotechnology have the potential to make valuable contributions to biology and medicine, which in turn could contribute to a new era in healthcare. The Federal agencies' R&D investments, for example, have resulted in advanced materials, tools, and nanotechnology-enabled instrumentation that can be used to study and understand biological processes in health and disease. The NIH-supported R&D efforts, in particular, are bringing about new paradigms in the detection, diagnosis, and treatment of common and rare diseases, resulting in new classes of nanotherapeutics and diagnostic biomarkers, tests, and devices.

The NIH became a participant in the NNI in 2001. The NNI serves as a framework within which the NIH can work collaboratively with other agencies to address some of the most perplexing challenges in the development and application of nanotechnologies for biomedical applications. Through this interagency planning, coordination, and communication, scientists are addressing key challenges by:

- Understanding the manner in which nanoscale building blocks and processes integrate and assemble into larger systems and how these processes can be precisely controlled to achieve predictable outcomes.

- Learning how to design nanomaterials that can seamlessly and functionally integrate with tissues of the body to perform biological functions.

- Developing "top-down" and "bottom-up" engineering approaches to control properties that allow the identification, characterization, and quantification of biological molecules, chemicals, and structures for early-stage changes or progression in a disease state.

- Engineering complex, theranostic-based nanoparticles and nanodevices to target therapies and diagnose the progress of treatments.

- Adopting new materials, nanotechnology-enabled tools, and analytical instruments from diverse fields of research.

The NIH continues to support the NNI by stimulating R&D in nanoscience and nanotechnology through both intramural and extramural funding activities in all eight program component areas, with major financial investments in fundamental nanoscale phenomena and processes (PCA 1), nanomaterials (PCA 2), and nanoscale devices and systems (PCA 3). For more information on specific topics funded by the NIH, please visit the NIH Research Portfolio Online Reporting Tool at http://report.nih.gov. The NIH plays a substantive role in developing scientific understanding of how to design nanomaterials for safe use in manufacturing and in medical treatments. The National Cancer Institute (NCI), for example, established the Nanotechnology Characterization Laboratory, which has developed a comprehensive assay portfolio for the assessment of the safety of nanoparticles in *in vivo* applications, and the National Institute of Environmental Health Sciences (NIEHS) and the National Toxicology Program have focused on assessing properties relevant to the chronic exposure of workers to nanomaterials. The NIH institutes also support large centers grants, program grants, and small businesses whose technologies or products are licensed or currently undergoing Phase I–III clinical trials.

National Institute for Occupational Safety and Health

The National Institute for Occupational Safety and Health (NIOSH) is responsible for conducting research and providing guidance to protect the health and safety of people at work. Workers are generally the first people in society to be exposed to the hazards of an emerging technology, and nanotechnology is no exception. The workplaces where nanomaterials are developed, investigated, manufactured, used, and disposed of are quite varied and span all economic sectors. To protect the health and safety of workers in all these workplaces, NIOSH has mounted a concerted R&D and public outreach effort that includes hazard identification, exposure assessment, risk characterization, and risk management.

NIOSH toxicology studies have provided better understanding of the ways in which some types of nanoparticles may enter the body and interact with the body's organ systems; however, the breadth and depth of such research efforts have been limited to a few nanoparticle types. More types of engineered nanoparticles need to be assessed for characteristics and properties relevant for predicting potential health risks. The toxicology studies will serve as a starting point to identify the priority materials for further risk assessment, exposure evaluations, and risk management practices.

NIOSH field investigators have assessed exposure to engineered nanoparticles in a limited number of workplaces, but few data exist on the full extent and magnitude of workers' exposures to broad categories of nanoparticles in workplaces that manufacture or use nanomaterials, nanostructures, and nanodevices. This effort will allow NIOSH field investigators to expand the scope of assessment and the number and type of facilities that can be assessed.

NIOSH guidance is a first step toward controlling nanoparticles in the workplace; however, more research is needed on the efficacy and specificity of engineering and work-practice control measures. Significantly more field research is needed to develop guidance, based on evaluating possible short- and long-term health risks in nanotechnology workers, and to develop guidance for medical surveillance and prospective epidemiologic studies.

NIOSH will continue to work with the NNI and a broad range of national and international partners to develop research-based information and guidance to protect workers involved with nanomaterials. The results being produced by NIOSH will continue to serve as the foundation for meeting the critical NNI research needs related to human exposure assessment, exposure mitigation, risk assessment techniques, risk management practices, and human medical surveillance and epidemiology. NIOSH has developed formal collaborations with the National Toxicology Program (NIEHS), CPSC, OSHA, and DOD. It has also developed productive informal interactions with additional agencies, including EPA, NIST, DOE, and FDA.

National Institute of Standards and Technology

Advancing nanoscale measurement science, standards, and nanotechnology is an important component of the National Institute of Standards and Technology (NIST) mission to promote U.S. innovation and industrial competitiveness. From leading cutting-edge research to coordinating the development of standards that promote trade and enable regulation of nanotechnology-based products, NIST's nanotechnology program directly impacts priorities important to the nation's economy and well-being. The NNI-related research conducted in NIST's laboratories and user facilities develops measurements, standards, and data crucial to a wide range of industries and Federal agencies, from the development of new spectroscopic methods needed to increase efficiency in advanced photovoltaics, to the development of the reference materials and data necessary to accurately quantify and measure the presence and impact of nanomaterials in the environment. NIST further supports the U.S. nanotechnology enterprise from discovery to production through the Center for Nanoscale Science and Technology (CNST) user facility, created under the NNI as the only national nanotechnology center with a focus on commerce. The CNST provides industry, academia, NIST, and other Federal agencies with access to world-class nanoscale measurement and fabrication methods and technology. NIST also accelerates U.S. innovation in nanotechnology by funding high-risk, high-reward research through the Technology Innovation Program (TIP), including targeted investments in nanomanufacturing research.

The NNI has enabled NIST to prioritize and coordinate nanotechnology research in numerous areas, most notably in nanoelectronics, nanomanufacturing, energy, and environmental, health, and safety aspects of nanomaterials (nanoEHS). NIST is working closely with other NNI agencies in planning and implementing the nanotechnology signature initiatives related to nanomanufacturing and energy. Through activities of the NSET Subcommittee's Nanotechnology Environmental and Health Implications (NEHI) Working Group, NIST has received input from a broad range of stakeholders on the critical measurement science and measurement tools—protocols, standards, instruments, models, and validated data—required for risk assessment and management of engineered nanoscale materials and nanotechnology-based products. This input has been essential to the development of NIST's nanoEHS program, including planning goals and milestones.

NIST staff members participate widely in nanotechnology-related standards development and international cooperation activities in order to promote transfer of NIST research, technology, and measurement services, and to advance NNI objectives within the Department of Commerce mission. The development of nanotechnology standards and guidelines is conducted through international fora such as the International Organization for Standardization's Technical Committee 229, the International Electrotechnical Commission's Technical Committee 113, ASTM International's Committee E56, or the OECD's Working Party on Manufactured Nanomaterials, supported by NIST staff in important leadership roles and coordinated with other agencies through the Global Issues in Nanotechnology (GIN) Working Group.

National Science Foundation

The National Science Foundation (NSF) supports fundamental nanoscale science and engineering in and across all disciplines. It also advances nanotechnology innovation through a variety of translational research programs and by partnering with industry, states, and other agencies.

The NSF nanotechnology investment in 2010 supported over 4,500 active projects, over 30 research centers, and several infrastructure networks for device development, computation, and education. It impacted over 10,000 students and teachers. Approximately 150 small businesses have been funded to perform research and product development in nanotechnology through the Small Business Innovation Research (SBIR) and Small Business Technology Transfer (STTR) programs. NSF's nanotechnology research is supported primarily through grants to individuals, teams, and centers at U.S. academic institutions. The efforts in team and center projects have been particularly fruitful because nanoscale research and education are inherently interdisciplinary pursuits, often combining elements of materials science, engineering, chemistry, physics, and biology.

Fundamental changes envisioned through nanotechnology require a long-term R&D vision. NSF sponsored the first initiative dedicated to nanoparticles in 1991, the 1997–1999 program Partnership in Nanotechnology, and produced the 1999 interagency report *Nanotechnology Research Directions: Vision for Nanotechnology in the Next Decade*, adopted as an official NSTC document in 2000. NSF continues to push the frontiers of science and technology innovations through continual interaction with the nanotechnology community, new programs, and ongoing evaluation of current investments. The NSF-led study *Nanotechnology Research Directions for Societal Needs in 2020* was released in 2010 (see: http://wtec.org/nano2/). With input from academic, industry, and government experts from over 35 countries, the report addresses the progress and impact of nanotechnology since 2000 as well as the vision and research directions for nanotechnology in the next ten years.

NSF supports the three NNI nanotechnology signature initiatives through core programs and new solicitations. NSF requested additional funds in 2011 for nanomanufacturing to support new concepts for high-rate synthesis and processing of nanostructures, nanostructured catalysts, nanobiotechnology methods, and methods to fabricate devices, assemble them into systems, and then further assemble them into larger-scale structures of relevance to industry. Environmental, health, and safety implications of nanotechnology, including development of predictive toxicity of nanomaterials, will be investigated in three dedicated multidisciplinary centers and in over 60 other smaller groups.

NSF has a focus on addressing education and societal dimensions of nanotechnology. Education-related activities include development of materials for schools, curricula for nanoscience and engineering, new teaching tools, undergraduate programs, technical training, and public outreach programs. The Nanoscale Informal Science Education Network (NISEnet) is a national network for nanotechnology education public outreach supported by the NSF. Research directed at identifying and quantifying the broad implications of nanotechnology for society, including social, economic, workforce, educational, ethical, and legal implications, is investigated in small groups and in the Nanotechnology in Society Network.

Nuclear Regulatory Commission

The mission of the U.S. Nuclear Regulatory Commission (NRC) is to license and regulate the nation's civilian use of byproduct, source, and special nuclear materials in order to protect public health and safety, promote the common defense and security, and protect the environment. NRC's scope of responsibility includes regulation of commercial nuclear power plants; research, test, and training reactors; nuclear fuel cycle facilities; medical, academic, and industrial uses of radioactive materials; and transport, storage, and disposal of radioactive materials and waste. In addition, NRC licenses the import and export of radioactive materials and works to enhance nuclear safety and security throughout the world.

As a regulatory agency, NRC does not typically sponsor fundamental research or product development. Rather NRC is focused in part on confirmatory research to verify the safe application of new technologies in the civilian nuclear industry. Currently the agency's focus with respect to nanotechnology is to monitor developments that might be applied within the nuclear industry to help NRC carry out its oversight role.

U.S. Geological Survey, Department of the Interior

At the U.S. Geological Survey (USGS), nanotechnology research involves the effects of nanoparticles at various levels of biological organization, from the molecular to the ecosystem level (details at http://microbiology.usgs.gov/nanotechnology.html). Much of USGS nanotechnology research focuses on assessing the occurrence, fate, and effects of naturally-occurring and engineered chemical contaminants in aquatic environments, or research on methods of detection of metal nanomaterials. Several programs provide information on nanoparticles or other contaminants, including the Contaminant Biology Program (http://biology.usgs.gov/contaminant/), the Toxic Substances Hydrology Program (http://toxics.usgs.gov/), the National Research Program (http://water.usgs.gov/nrp/), and the Water Resources Research Institutes (http://water.usgs.gov/wrri/index.html). The NNI, through regular NSET Subcommittee meetings and activities within the NSET working groups, provides mechanisms for USGS to share information on nanotechnology research and to collaborate with other agencies.

The U.S. International Trade Commission

The U.S. International Trade Commission (USITC) is an observing member of the NSET Subcommittee. The USITC representative attends meetings to keep the Commission abreast of current trends and issues related to nanotechnology that may have the potential to impact international trade.

U.S. Patent and Trademark Office

The strength and vitality of the U.S. economy depends directly on effective mechanisms that protect new ideas and investments in innovation and creativity. The U.S. Patent and Trademark Office (USPTO) is at the cutting edge of the nation's technological progress and achievement as the Federal agency responsible for granting patents, registering trademarks, and providing intellectual property policy advice and guidance to the Executive Branch. Through its participation in the NNI, and working with other agencies through the NSET Subcommittee, the USPTO has made several improvements to its processes to keep pace with the rapid advances being made in this area. Notably, the USPTO adopted the NNI definition of nanotechnology in its development of the first detailed, patent-related nanotechnology classification hierarchy of any major intellectual property office in the world. The USPTO has also used the networking and information-sharing opportunities presented by participation in the NNI to establish nanotechnology-related training opportunities for patent examiners. The USPTO has significantly contributed to the NNI by providing advice on patent and other intellectual property-related matters, as well as contributing a variety of nanotechnology-related patent data, which have been used as a benchmark to analyze nanotechnology development and to perform trend analysis of nanotechnology patenting activity in the United States and globally.

Goals and Objectives

The NNI vision is supported by the four NNI goals. All four are equally critical to the success of the NNI and are interdependent. This interconnection is specifically recognized, as appropriate, in the following sections that describe shared NNI objectives, organized by NNI goal. Recurring themes that are particularly relevant for the realization of these objectives include the need for documentary standards; education and training; consideration of ethical, legal, and societal implications; public engagement; and environmental, health, and safety research. Based on extensive external and stakeholder input, the NNI agencies have assembled the objectives that follow. To the extent possible, these are specific and measurable, with targeted time frames of three to five years unless otherwise indicated. Although not all member agencies are responsible for fulfilling all objectives, the NSET Subcommittee has identified objectives that are supported by the relevant agencies and are both realistic and extending. The Subcommittee has attempted to recognize available resources and functional limits while also being far-sighted in terms of accelerating innovation and progress toward achieving the NNI goals. NNI agencies also independently continue to contribute to the achievement of all four goals through a number of their own activities, which are reported on an annual basis in the NNI supplement to the President's Budget.[8]

The actions and associated resources required to implement the goals of this plan will need to be prioritized in the context of other U.S. Government priorities. This document is not intended as an inherent justification to seek increased budgetary authority. The goals and objectives may be achieved through reprioritization and reallocation of existing resources. It is expected that departments and agencies will consider this document in their internal prioritization and planning processes.

Goal 1: Advance a world-class nanotechnology research and development program.

The NNI continues to expand the boundaries of knowledge and develop technologies through comprehensive and focused R&D within the participating agencies. The overarching objective of Goal 1 is to advance nanoscience and nanotechnology through the implementation of the objectives described below. Progress in R&D will depend upon the availability of a skilled workforce, infrastructure, and tools (Goal 3) and will lay the foundation for responsible incorporation of nanotechnology into commercial products (Goals 2 and 4).

Goal 1 Objectives

1.1 Continue to support R&D at the frontiers and intersections of scientific disciplines in the form of intramural and extramural programs targeting single investigators, multi-investigator and multidisciplinary research teams, and centers for focused research.

The broad NNI R&D portfolio invests at the frontiers and intersections of many disciplines, including biology, chemistry, ecology, engineering, geology, materials science, medicine, physics, and social sciences. Activities targeted toward this goal include support for fundamental research, use-inspired research, applications

---

[8] Available at http://nano.gov.

research, and technology development. The research efforts of the NNI agencies continue to be executed through a balanced mix of funding ranging from single-investigator grants to research centers and user facilities, each of which plays a unique and vital role in the discovery and innovation process.

1.2 Develop at least five broad interdisciplinary nanotechnology initiatives that are each supported by three or more NNI member agencies and support significant national priorities.

No single agency within the Federal Government has the mission or breadth of expertise to fully exploit the opportunities presented by nanotechnology, or to execute all the requisite research. Thus, in certain key areas, it is essential to coordinate particular NNI R&D programs across multiple agencies. NNI member agencies will identify topical areas that can most benefit from close and targeted interagency interaction, or nanotechnology signature initiatives. These will be implemented through the broad range of funding mechanisms identified in Objective 1.1 and will be coordinated to foster innovation and accelerate nanotechnology development.

1.3 Identify and support goal-oriented nanoscale science and technology research aimed at national priorities informed by active engagement with academia, industry, and other stakeholders.

Successful commercialization of nanotechnology (Goal 2) will depend on the scientific quality of research; better understanding of the potential environmental, health, and safety implications of nanotechnology; and cognizance of its relevance and competitiveness in the marketplace. The NNI member agencies will continue to work with academia and across industry sectors to gather input and feedback on Federal research. This continuous engagement will facilitate the effective transition of nanotechnology from discovery to the marketplace. Such engagement could be fostered via means such as matching funds, partnerships, consortia, and planning exercises.

1.4 Develop quantitative measures to assess the performance of the U.S. nanotechnology R&D program relative to that of other major economies, in coordination with broader efforts to develop metrics for innovation.

Nanotechnology is a worldwide field with significant R&D efforts underway in many countries. In order to maintain U.S. leadership, it is critical to develop clearly defined metrics to measure the U.S. R&D program against those of other major economies. Efforts to measure innovation are already underway in other areas,[9] and the NNI will leverage that existing work.

Goal 2: Foster the transfer of new technologies into products for commercial and public benefit.

Significant advances have been made in the fundamental understanding of nanotechnology over the past ten years. While nanotechnology has found its way into commercial products, e.g., in the areas of cosmetics, electronics, and healthcare, a continued emphasis on commercialization is essential to fully realizing the benefits of nanotechnology R&D to the nation. The purpose of Goal 2 is to establish

---

[9] See, for example, the interagency STAR METRICS Project, an effort to track science investments and innovation (http://scienceofsciencepolicy.net/), and international projects such as the OECD Innovation Strategy, available at http://www.oecd.org.

processes to facilitate the responsible (Goal 4) transfer of nanotechnology research (Goal 1) into practical applications and capture its benefits to national security, economic development, and job creation (Goal 3). Successful completion of the objectives of Goal 2 requires close coordination with progress toward the other three NNI goals.

Several factors are necessary to achieve the successful commercialization of any new technology. Scalable, repeatable, cost-effective manufacturing methods are required to move the technology from the laboratory into commercial products. Both public and private sector investments are needed to help technologies reach maturity. Maximizing the benefits of nanotechnology developments to the U.S. economy also requires efforts to remove barriers to global commercialization and an understanding of the potential markets for a given product.

The NNI fosters technology transfer by facilitating agency engagement with key industry sectors to understand their technology needs, by providing industry and the public with access to the results of federally funded nanotechnology research, and by helping to support the creation of a business environment conducive to responsible development of nanotechnology. Partners in this undertaking include international, regional, state, and local organizations that promote nanotechnology development, as well as professional societies, trade associations, and other nongovernmental organizations.

Goal 2 Objectives

2.1 Develop robust, scalable nanomanufacturing methods necessary to facilitate commercialization by doubling the share of the NNI investment in nanomanufacturing research over the next five years.

Nanomanufacturing R&D involves a fundamental understanding of the manufacturing process, including the development and application of measurement and characterization techniques, reference materials, and standards. The 2010 review of the NNI by the President's Council of Advisors on Science and Technology (PCAST)[10] recommended a greater emphasis on commercialization by doubling the investment of the Federal Government in nanomanufacturing. In addition, the PCAST report recommended the initiation of interagency partnerships, i.e., signature initiatives, within the next 3 years. Along these lines, a nanotechnology signature initiative in sustainable nanomanufacturing is planned for initiation in FY 2011.

2.2 Increase focus on nanotechnology-based commercialization and related support for public-private partnerships, by:

> 2.2.1. Launching at least five public-private partnerships over the next five years.

> 2.2.2. Leveraging the NNI nanotechnology signature initiatives (see Goal 1 objectives) to remove barriers to commercialization of nanotechnology innovations, particularly in areas of high national need.

> 2.2.3. Working with U.S. industry across sectors to develop technology "roadmaps" or long-term R&D plans, as appropriate, in support of new public-private partnerships and signature initiatives.

---

[10] See Appendix A.

Many nanotechnology products are moving into commercialization phases, and some NNI member agencies are increasingly collaborating with diverse industry sectors as products are brought to market. The NNI fosters responsible technology transfer through the NSET Subcommittee and its member agencies engaging with key industry sectors and collecting and exchanging information and ideas regarding each sector's technology needs. It also provides a bridge between companies and federally funded nanotechnology research, strengthening Goal 1 outcomes. This collaborative work could be a market driver, potentially enhancing the U.S. economy and job creation. Such cooperative efforts will be strengthened by supporting public-private partnerships and by an ongoing effort to work with industry as the technology matures. This effort can also help mission-oriented agencies' efforts to expand their applied research and commercialization programs. In the future, the NSET Subcommittee will explore the feasibility of additional mechanisms such as innovation clusters, matching funds, and other collaborative efforts to facilitate the commercialization of emerging technologies.

2.3 Establish and/or sustain national user facilities, cooperative research centers, and regional initiatives with the goal of accelerating the transfer of nanoscale science from discovery to commercial products, by:

> 2.3.1. Providing economical access by academia and industry, on both precompetitive and proprietary bases, to state-of-the-art tools and processes, expertise, and training that are critical to the transition from discovery to advanced prototype, with options for remote use when feasible.

> 2.3.2. Supporting the establishment over the next five years of at least three self-sustaining cooperative research centers and/or state and regional economic development initiatives for nanotechnology.

Over the past decade, NNI member agencies have made considerable investments in the development of unique national facilities to support nanotechnology R&D. These investments maintain the existing infrastructures as well as capabilities needed to support both basic research in nanotechnology and commercialization efforts. Further efforts to promote nanotechnology commercialization can be supported through continuing the development of government-university-industry consortia and economic development initiatives at the state or regional level. The NNI will continue to coordinate with regional, state, and local nanotechnology initiatives through activities such as NNI workshops[11], webinars, and other events that provide a forum for communication and collaboration in this community.

2.4 Assist the nanotechnology-based business community, including small- and medium-sized enterprises, in understanding the Federal Government's R&D funding and regulatory environment, by:

> 2.4.1. Developing and disseminating informational materials documenting funding opportunities (e.g., in the SBIR and STTR programs), state-of-the-art nanotechnology user facilities that are available to industry, and other resources available from the Federal Government.

---

[11] See http://nano.gov for more information on NNI workshops with regional, state, and local nanotechnology initiatives.

2.4.2. Providing informational materials, including points of contact, to explain issues such as environmental, health, and safety regulations and export controls pertinent to nanotechnology-related products and businesses.

NSET member agencies recognize the need to make the business community aware of Federal Government resources that are available for helping foster nanotechnology-based commercialization and economic development efforts, and of the Federal regulations that may apply to these efforts. Small and medium-sized businesses in particular may not have dedicated staff with specialized expertise related to Federal resources and regulations.

2.5 Increase international engagement to facilitate the responsible and sustainable commercialization, technology transfer, innovation, and trade related to nanotechnology-enabled products and processes, by:

2.5.1. Increasing the participation of NNI member agencies, when appropriate, in fora addressing global legal, intellectual property, and regulatory issues related to nanotechnology-enabled product development.

2.5.2. Enhancing interagency communication and collaboration towards assuring safe nanotechnology-enabled products for domestic and international consumers, through activities such as developing documentary standards.

The successful commercialization of nanotechnology products in domestic and global markets is dependent on adequately addressing issues such as intellectual property (IP), return on investment, and environmental, health, and safety regulations and concerns. NNI member agencies' continued involvement in activities related to these issues is needed not only to ensure a safe environment but also to maintain a level playing field for all involved. For example, ongoing interagency support of development of U.S. and international documentary standards will facilitate such innovation and product development. Successful commercialization also involves the purposeful integration of the aims of each of the other goals in areas such as safeguarding research and IP investment (Goal 1), maintaining a highly skilled workforce (Goal 3), and ensuring responsible and sustainable development, including environmental benefits (Goal 4).

Goal 3: Develop and sustain educational resources, a skilled workforce, and the supporting infrastructure and tools to advance nanotechnology.

Fundamental to the continued successful development of nanotechnology is the development of the infrastructure necessary to support this effort. A substantial investment, strengthened by and dependent on interagency cooperation and collaboration through the NNI, is needed to develop the talent and resources necessary to achieve the other NNI goals of advancing a world-class R&D program (Goal 1), fostering the transfer of new technologies into products for commercial and public benefit (Goal 2), and supporting responsible development of nanotechnology (Goal 4).

Nanotechnology is emerging amid a transformative phase in education in the United States when there is a widely recognized need to improve science, technology, engineering, and mathematics (STEM) education. The creation in the United States of a world-leading science and technology workforce, including careers in biomedical science, can be accelerated by nurturing students' interest in STEM topics. Not only do core science concepts underpin nanoscale science and engineering, but the discovery of emergent properties and behaviors at the nanoscale can create a "wow" factor to inspire students to learn about nanotechnology and STEM more broadly. Innovations in nanotechnology can be exploited as vehicles for learning and teaching STEM subjects that students have traditionally found to be too theoretical or too challenging.

The NNI continues to foster educational programs that develop scientists, engineers, technicians, production workers, and laboratory personnel (including academic students and trainees) through multidisciplinary academic programs, industrial partnerships and federally funded R&D systems. Extensive infrastructure capabilities, which include the centers and user facilities supporting research on nanomanufacturing, nanoscale characterization, synthesis, simulation, and modeling that have been developed through the NNI over the past ten years, will continue to be advanced. For the NNI to meet the goals outlined in this plan, it will be essential in the coming years to connect and coordinate the vast amount of information pertaining to nanotechnology.

Goal 3 Objectives

3.1 Initiate, develop, support, and sustain programs for educating, training, and maintaining a skilled nanotechnology workforce.

The demand for technicians and research scientists to work in nanotechnology-related industries is anticipated to increase with the maturation of a number of nanotechnology-enabled products and processes. With the support of NNI centers, colleges and universities have been offering undergraduate minors and majors, teacher training, and postgraduate programs in nanoscale science and engineering. In order to prepare high school graduates for careers in nanotechnology-related industries, the NNI member agencies will work collaboratively to support the development of K–12 STEM (and related, including biomedical) curriculum standards and articulation plans that incorporate problem-based and integrative teaching, where appropriate. International standards and best practices (e.g., for safe handling of nanomaterials in the laboratory, as described in Goal 4) will help to inform these developments. Information on nanotechnology and nanoscience-based career opportunities and workforce needs will strengthen the pursuit of this objective. Online resources should be utilized to help disseminate information about nanotechnology careers and formal education programs in nanotechnology.

3.2 Initiate outreach and informal education programs and publish related information to foster a student population, workforce, and public that are well informed about the opportunities in nanotechnology-related industries and the potential impacts of environmental, health, and safety (EHS) and ethical, legal, and societal implications (ELSI) of nanotechnology.

The information technology (IT) revolution reached the public through its integration into virtually all aspects of our lives. Whereas "IT" has become a commonplace term associated with specific applications, the technology behind nanotechnology-enabled products may result in tremendous enhancements or entirely new product properties that might not be explicitly referred to as "nano." Multiple communication tools (e.g., print media, online webcasts and podcasts, museum exhibits, and special events) will be used to achieve this objective.

3.3 Provide, facilitate the sharing of, and sustain the physical R&D infrastructure for nanoscale fabrication, synthesis, characterization, modeling, design, computation, and hands-on training for use by industry, academia, nonprofit organizations, and state and Federal agencies, by:

> 3.3.1. Determining the current capacity and inventory of tools, facilities and supporting infrastructure, and staffing and services that are available, and determining the capacity requirements up to the year 2020.

> 3.3.2. Developing, operating, maintaining, and sustaining highly advanced tools, infrastructure, and user facilities (including investment, staffing, and upgrades).

Robust nanotechnology R&D and technical advancement will require the support of a state-of-the-art physical infrastructure that is widely accessible. The specialized capability, equipment, and structures needed for nanoscience R&D are prohibitively expensive for small enterprises and educational institutions. Sustained and predictable access to a broad range of state-of-the-art instrumentation and facilities for synthesis, processing, fabrication, characterization, modeling, and analysis of nanomaterials and nanosystems, including bio-nanosystems, is needed to achieve this objective. In most cases, no single researcher or even single institution can justify funding the acquisition of and support for all necessary tools, and therefore user facilities that provide access to researchers from multiple sectors, including academia and industry, serve a critical role. Such facilities have the ability to co-locate a broad suite of necessary nanotechnology tools, to maintain these tools at the leading edge, and to provide staff with expertise to ensure the most productive use of the tools. In addition, they provide an outstanding setting for hands-on training of nanotechnology researchers.

The extensive infrastructure established by the NNI over the past ten years will be upgraded and sustained based on evaluations of the need and capacity requirements. International best practices will be incorporated into the current infrastructure, as appropriate. Extensive publicity and dissemination of information will help to reach the nanotechnology sector, especially small and medium enterprises, to ensure that this infrastructure is accessible to all and well utilized.

Goal 4: Support responsible development of nanotechnology.

Responsible development of nanotechnology is central to advancing a world-class R&D program (Goal 1), educating the workforce and engaging the public (Goal 3), and all aspects of nanomanufacturing and product commercialization (Goal 2). To help integrate responsible development across the spectrum of nanotechnology, the NNI has developed, with input from stakeholders, a nanotechnology-related environmental, health, and safety (EHS) research strategy with a broad, multi-agency perspective.[12] Research in support of Goal 4 provides information and data for research institutions, regulatory agencies, the public, and industry, helping to assure that nanotechnology-enabled products minimize adverse impacts and maximize benefits to humans and the environment.

The Goal 4 objectives facilitate and track progress in responsible development and are divided into four different yet integral sections: public health and environmental protection; domestic and international engagement; ethical, legal, and societal implications of nanotechnology; and solutions for critical public health and environmental challenges.

Goal 4 Objectives

4.1 Incorporate safety evaluation of nanomaterials[13] into the product life cycle, foster responsible development, and where appropriate, sustainability across the nanotechnology innovation pipeline, by:

> 4.1.1. Developing and applying:
>
>> 4.1.1.1. Measurement and screening tools (defined as protocols, standards, models, data, and instruments) to assess the physico-chemical properties of nanomaterials and their biological effects in the environment and on human health and to quantify exposure across the nanotechnology product life cycle.
>>
>> 4.1.1.2. Models, including risk assessment models, to assess safety of nanomaterials throughout the life cycle of the material or product.
>>
>> 4.1.1.3. Health surveillance models as appropriate for the nanotechnology workforce (including laboratory personnel such as students, technicians, and trainees), consumers, susceptible populations, and the environment.
>
> 4.1.2. Creating mechanisms for appropriate and timely information sharing and dissemination among stakeholders including academia, industry, legal entities, Federal agencies, regulatory communities, governments (e.g., state, local, and tribal), the general public, and other relevant stakeholders.
>
> 4.1.3. Establishing guidance, standards, or other methods to formulate nanotechnology-related regulatory approaches for domestic and global researchers, manufacturers, distributors, and users of nanotechnology-enabled products to ensure the protection of public health and the environment.

In support of this objective, the NNI Environmental, Health, and Safety Research Strategy[14] ("NNI EHS Research Strategy") provides a research framework, including exposure and hazard identification across

---

[12] See the *NNI Environmental Health and Safety Research Strategy*, available in early 2011 at http://nano.gov. This strategy, informed by a series of four EHS-themed stakeholder workshops in 2009–2010, consists of comprehensive planning and research coordination, as well as a review of research needs and accomplishments.

[13] Consistent with other NNI EHS documents, the term *nanomaterials* here refers to engineered nanoscale materials, i.e., those materials that have been purposely synthesized or manufactured to have at least one external dimension of approximately 1–100 nanometers—at the nanoscale—and that exhibit unique properties determined by this size. Nanotechnology-enabled products encompass intermediate products that exist during manufacture as well as final products.

[14] Available in early 2011 at http://nano.gov.

the nanomaterial and product life cycle, by identifying core research needs in the areas of human exposure, the environment, human health, and measurement tools. The NNI EHS Research Strategy is complemented by risk assessment and risk management approaches along with research needs in predictive modeling. It serves as guidance to Federal agencies as they develop their agency-specific nanotechnology EHS strategies and implementation plans.

4.2 Develop tools and procedures for domestic and international outreach and engagement to assist stakeholders in developing best practices for communicating and managing risk, by:

4.2.1. Identifying information gaps and prioritizing research essential for risk communication and risk management, which can address potential occupational and product hazards for start-up and larger companies working with nanomaterials and nanoscale processes, to:

4.2.1.1. Assure adequacy of workforce training and risk communication strategies through active outreach and engagement.

4.2.1.2. Increase available information for better decision making in assessing and managing risks from nanomaterials.

4.2.2. Obtaining stakeholder perspectives by developing and using a variety of methods, such as surveys, workshops, public meetings, and advisory panels; disseminating information through publicly accessible summaries of findings; and developing mechanisms for integration of EHS priorities and assessment methods into national and international regulatory policies.

4.2.3. Communicating available information about assessing and managing potential risks from nanomaterials and about nanotechnology-related regulatory approaches to both domestic and global manufacturers.

4.2.4. Increasing U.S. participation internationally in bilateral and multilateral fora and organizations that address stakeholders' concerns surrounding the development of nanotechnology by providing information, guidance, training, and capacity-building resources for governments.

The NNI EHS Research Strategy provides details on how the NNI can better develop and disseminate knowledge with a range of stakeholders, and describes NNI efforts to engage internationally in the area of EHS research. Federal agencies actively engage with other countries on a bilateral and multilateral basis to help further this objective.

4.3 Identify and manage the ethical, legal, and societal implications (ELSI) of research leading to nanotechnology-enabled products and processes, by:

4.3.1. Increasing the capacity of Federal agencies to identify and address ELSI issues specific to nanotechnology by fostering the development of a community of expertise on ELSI issues related to nanotechnology and by providing a current resource list of experts that is accessible to a broad range of users.

4.3.2. Building collaborations among the relevant communities (e.g., consumers, engineers, ethicists, manufacturers, nongovernmental organizations, regulators, and scientists—including social and behavioral scientists) to enable prompt consideration of the potential risks and benefits of research breakthroughs and to provide perspectives on new research directions.

4.3.3. Developing information resources for ethical and legal issues related to intellectual property (IP) and ethical implications of nanotechnology-based patents and trade secrets.

ELSI issues are interwoven with all of the NNI goals and are integrated into each of the research needs described in the NNI EHS Research Strategy to help Federal agencies consider stakeholder concerns when identifying research areas and establishing decision analysis methodologies. The resources embodied by these objectives will help agencies to develop more robust nanotechnology-related ELSI research portfolios attuned to their missions.

4.4 Employ nanotechnology and sustainable best practices to protect and improve human health and the environment, by:

4.4.1. Supporting research to incorporate environmentally benign methods into manufacturing processes.

4.4.2. Developing technologies to assess the status of human health and ecosystems.

4.4.3. Fostering the use of nanomaterials as safer substitutes for commonly used compounds that have known adverse effects on human health and the environment.

4.4.4. Creating and implementing methods, nanomaterials, and nanotechnology-enabled devices to reduce human and environmental exposures to harmful compounds and remediate environmental contamination.

Nanotechnology can play a role in solving societal challenges such as access to safe food and water, secure living and work environments, clean and renewable energy, and diagnosis and treatment of diseases or medical disorders. Research directed at applications is complementary to EHS research; the support of both is needed to realize the NNI goal of responsible nanotechnology development.

Coordination and Assessment

The NNI is coordinated, planned, implemented, and reviewed by the Nanoscale Science, Engineering, and Technology (NSET) Subcommittee of the Committee on Technology (CoT) of the National Science and Technology Council (NSTC). Other components of NNI coordination include thematic NSET working groups (described below), the National Nanotechnology Coordination Office (NNCO), and the Executive Office of the President (EOP). Periodic assessment of the NNI by external advisory bodies provides additional input and guidance to the NNI. Figure 2 shows the relationships between NNI coordination and assessment bodies. The roles of these entities are further described below.

Nanoscale Science, Engineering, and Technology Subcommittee

The Nanoscale Science, Engineering, and Technology (NSET) Subcommittee was established in 2000 under the NSTC's Committee on Technology (CoT) to coordinate interagency nanotechnology R&D activities. The National Nanotechnology Coordination Office (NNCO) was subsequently established as the point of contact on Federal nanotechnology R&D activities and to provide technical and administrative assistance to the NSET Subcommittee. The 21st Century Nanotechnology Research and Development Act in December 2003 (hereafter referred to as "the Act") formalized many of the coordination structures that the NSTC had organized, and established additional mechanisms to ensure that the Federal Government developed sound, informed nanotechnology R&D strategies and policies. This legislation also created the National Nanotechnology Advisory Panel (NNAP), called for a triennial review of the NNI by the National Research Council of the National Academies (NRC/NA), and established functions for the NNCO.

The NSET Subcommittee leads the interagency coordination of the Federal Government's nanotechnology R&D enterprise by cooperatively coordinating the research, development, communication, and funding functions of the NNI. The NSET Subcommittee develops the NNI Strategic Plan, prepares the annual NNI supplement to the President's Budget, and sponsors workshops or other interagency activities that inform the Federal Government's nanotechnology-related decision-making processes.[15] The high level framework provided by the NNI Strategic Plan establishes goals, objectives, and priorities. It guides and informs the participating agencies in developing their nanotechnology R&D implementation plans. The subcommittee promotes balanced investment across all of the agencies to address the critical elements needed to support the development and utilization of nanotechnology. Further, the subcommittee interacts with pertinent academic, industry, state, and local government groups, and with international organizations. Each agency participating in the NNI is represented on the NSET Subcommittee; a list of those agencies is given at the front of this report. A co-chair from the Office of Science and Technology Policy (OSTP) and a co-chair from an NNI agency lead the NSET Subcommittee, which meets at least six times each year.

---

[15] Access to information on all NNI-sponsored workshops and complete versions of all reports and related NNI documents are available at http://nano.gov.

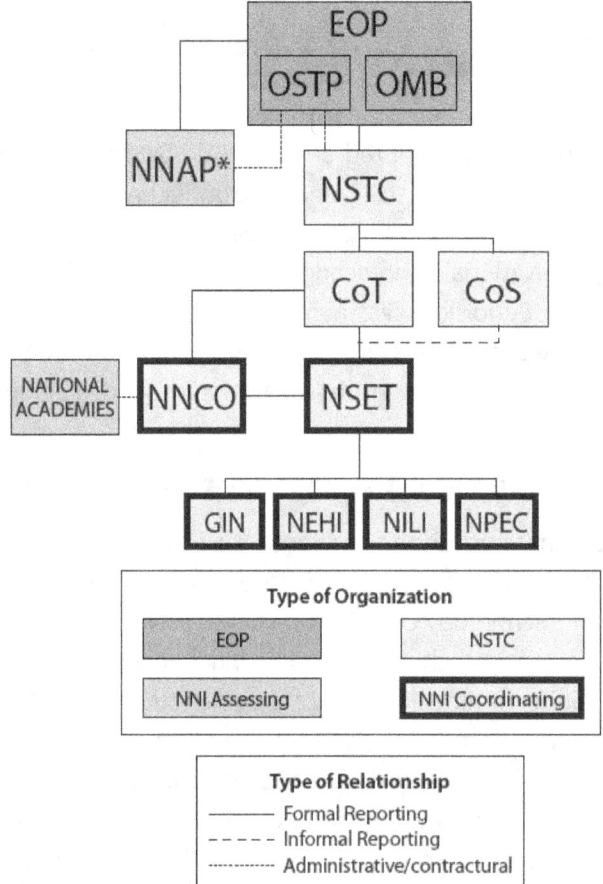

Figure 2. Coordination and Assessment of the NNI

*Executive order 13349 designates the President's Council of Advisors on
Science and Technology (PCAST) as the National Nanotechnology Advisory Panel (NNAP).*

Working Groups of the NSET Subcommittee

The NSET Subcommittee has chartered four working groups[16] to provide a structure to improve the effectiveness and productivity of the subcommittee and its member agencies in areas that will benefit from focused interagency attention and activity. These groups are Global Issues in Nanotechnology (GIN); Nanotechnology Environmental and Health Implications (NEHI); Nanomanufacturing, Industry Liaison, and Innovation (NILI); and Nanotechnology Public Engagement and Communications (NPEC). The NSET Subcommittee periodically reviews the need for existing or new working groups in terms of focus, intended participation, and scope, as reflected in the groups' charters.

Global Issues in Nanotechnology (GIN) Working Group

Global issues are interwoven among each of the NNI's four goals. Member agencies routinely engage with their international partners on a wide breadth of topics, from joint research programs to standards development to worker safety. The NSET Subcommittee's GIN Working Group strives to coordinate and

---

[16] The latest information on each of the NSET working groups is at http://nano.gov.

focus these activities in a manner consistent with the NNI vision and U.S. policy. GIN supports each of the four NNI goals by strengthening international R&D collaboration, capacity building, and engagement on regulatory and trade issues, all of which are essential to the development of a vibrant and safe global marketplace for nanomaterials and nanotechnology-enabled products. The working group also serves as the coordinating body for Federal Government activities in the OECD's Working Party on Nanotechnology (WPN), a leading intergovernmental forum that advises upon emerging policy issues of science, technology, and innovation related to the responsible development of nanotechnology.

Nanotechnology Environmental and Health Implications (NEHI) Working Group

Nanotechnology has the potential to significantly transform society in many key areas including new materials, processes, and products. In order to fully realize the promise of nanotechnology, Federal agencies support research to understand the environmental, health, and safety (EHS) implications of nanotechnology and provide guidance on the safety of nanomaterials across the product life cycle. The NSET Subcommittee's NEHI Working Group provides a forum for focused interagency collaborations on EHS and leadership in establishing the national nanotechnology EHS research agenda, in addition to communicating EHS information between NNI agencies and to the public. The combined efforts of the nanotechnology R&D community, public health advocacy groups, and the public are required to fully address EHS research priorities and strengthen the scientific foundation of risk assessment and risk management of nanotechnology. NEHI provides the nexus, as appropriate, for interactions between agencies and these diverse communities. The sum of these interactions and activities enhances the value of NNI efforts and provides a collaborative approach to examining public health and environmental concerns about nanomaterials. NEHI member agencies include those with direct responsibilities for public, workplace, and environmental safety, and agencies involved in science, education, and policy.

Nanomanufacturing, Industry Liaison, and Innovation (NILI) Working Group

A sustained commitment to nanotechnology-based innovation, made possible by dialogue and partnerships among all the players in the innovation ecosystem, is key to realizing the NNI vision. The NSET Subcommittee's NILI Working Group coordinates many activities in this area. NILI promotes and facilitates exchange of information among Federal agencies, academia, industry, and state, regional, and local organizations to build U.S. leadership in nanotechnology-based products and manufacturing processes through activities such as developing a database of nanotechnology-related technology transfer and nanomanufacturing programs across agencies; periodically organizing workshops that bring together regional, state, and local stakeholders; and supporting the NNI nanotechnology signature initiative on sustainable nanomanufacturing. In particular, NILI seeks to help the NNI agencies organize coherent support structures and effective technology transfer practices, making good use of the expertise of industry-initiated liaison groups. These liaison groups are a conduit for partnerships between the NNI and industry sectors and generally represent particular industries (e.g., electronics, chemicals, and forest products). A significant component of supporting the innovation process is fostering development of standard nanotechnology reference materials, terminology, and measurement and characterization methods. As with the other NSET Subcommittee working groups, communication and engagement activities are a significant focus of NILI.

Nanotechnology Public Engagement and Communications (NPEC) Working Group

Although public engagement and communication activities are critical to each NSET Working Group achieving its primary goals in support of the NNI vision, there remains a need for focused efforts to understand and promote best practices in public engagement and communication activities related to nanotechnology. The NSET Subcommittee's NPEC working group provides this focus. NPEC promotes and coordinates the efforts of NNI member agencies to educate and involve the public, policymakers, and stakeholder groups in discussions about nanotechnology, its applications and implications, the work of the NNI, and related topics of special interest. NPEC also assists in the development of research-based guidance for outreach and engagement among governmental and nongovernmental organizations, the public, and other stakeholders regarding the responsible development of nanotechnology.

National Nanotechnology Coordination Office (NNCO)

The National Nanotechnology Coordination Office (NNCO) serves as a pivotal locus for NNI activity, by providing technical and administrative support for the NSET Subcommittee; serving as a central point of contact for Federal nanotechnology R&D activities, including the NSET working groups; and performing public outreach and engagement on behalf of the NNI.

NNCO organizes meetings of the NSET Subcommittee and its working groups, providing staff members to serve as central points of contact and to record and maintain minutes of the meetings. NNCO also organizes NNI-sponsored workshops and prepares and publishes reports of those workshops. It coordinates the preparation and publication of NNI interagency planning, budget, and assessment documents, such as the annual NNI Supplement to the President's Budget. NNCO serves as a Congressional liaison by coordinating the development of information on the NNI and its activities for Congress when requested.

NNCO produces and distributes information for the general public, including brochures, workshop reports, nanotechnology-related news, educational resources, funding opportunities, and other information, all of which are made available on the NNI website, http://nano.gov. This website, which is designed, organized, and maintained by the NNCO, also provides information about recent developments in nanotechnology and NNI activities. The NNCO communications effort is strengthened by relationships between NNCO staff and key press contacts and public information officers at NSET member agencies. NNCO staff members prepare and deliver presentations and lectures on NNI activities at professional society meetings and at a wide variety of public venues. NNCO will continue to organize diverse public input and outreach activities; future examples may include interactive web dialogues, citizen panels, workshops, and other educational events.

Contributions from the NSET Subcommittee member agencies fund the NNCO. The White House Co-Chair of the NSTC Committee on Technology appoints the NNCO Director from a Federal agency, in consultation with the co-chairs of the NSET Subcommittee. The NNCO Director currently serves as the Coordinator for Standards Development. The NNCO Deputy Director is also detailed from a Federal agency and has been named as the Coordinator for Environmental, Health, and Safety Research.

Executive Office of the President

Representatives from the Executive Office of the President (EOP) participate in NNI activities to ensure that implementation of the NNI is coordinated and consistent with government-wide priorities. The primary points of interaction are the Office of Science and Technology Policy (OSTP) and the Office of Management and Budget (OMB).

OSTP is responsible for advising the EOP on matters relating to science and technology and supports coordination of interagency science and technology activities. OSTP administers the NSTC, and the OSTP representative to the NSET Subcommittee is a co-chair of the subcommittee. This arrangement provides EOP-level input on and support for various NNI activities.

OMB is responsible for coordinating with the NNI member agency budget offices to establish the nano-technology R&D budget for planning and tracking purposes. Each year, OMB collects budget information regarding the total Federal investment in nanotechnology R&D, as well as information about agency investments within each program component area.

Assessment

The Act calls for periodic assessment of the NNI through annual interagency reporting and review by external advisory bodies. The annual interagency analysis of progress called for in the Act is provided in the NNI supplement to the President's Budget, which also serves as the NNI annual report called for in the Act. Specifically, progress towards achieving NNI goals and priorities is analyzed in terms of (1) invest-ments categorized by PCA, including cross-cutting interagency activities coordinated through the NSET Subcommittee, and (2) activities relating to the four NNI goals, including individual agency activities as well as activities coordinated with other agencies and institutions, including international interactions.

Review by outside advisory groups is vital to keeping NNI efforts focused and balanced, and the Act established two mechanisms for such review. First, the Act calls for the President to establish a National Nanotechnology Advisory Panel (NNAP) to advise the President and the NSTC on matters relating to the NNI. The Act specifically calls for the NNAP to assess the Federal nanotechnology R&D program at least once every two years. Executive Order 13349 (2004) designates the President's Council of Advisors on Science and Technology (PCAST) as the NNAP. The members of PCAST are senior representatives from industry and academic research institutions who have extensive experience in managing large science and technology organizations. Second, the Act calls for the NNCO Director to make arrangements for the National Research Council of the National Academies (NRC/NA) to review the NNI every third year. NRC/NA panels for the NNI reviews are typically comprised of a broad cross-section of technical experts with knowledge specifically related to nanotechnology. The NRC/NA provides independent science, technology, and health policy advice to the Federal Government. It is the principal operating agency of the National Academies in providing services to the Federal Government, the public, and the scientific and engineering communities.

The first assessment by PCAST in its role as the NNAP was released in May 2005, and the first NRC/NA review under the Act was completed in November 2006. Subsequent reviews from PCAST were com-pleted in April 2008 and March 2010. The NRC/NA delivered its second triennial report assessing the

Federal strategy for nanotechnology-related environmental, health, and safety research in December 2009. The perspectives of these two bodies, and their assessments, are complementary, and the NNI has benefited from their diverse inputs into the planning and evaluation process. The resulting recommendations have led to specific actions and focused attention in areas that were highlighted by both groups, including research on environmental, health, and safety aspects of nanotechnology and expanded efforts to improve education and workforce preparation as well as program management.

The Path Forward

The full realization of the promise of nanotechnology as envisioned by the NNI and its participating agencies will continue to rely on the hallmark activities of the NNI, namely, to nurture coordination, collaboration, and communication among participating Federal agencies. In this area, means for improved interagency communication across all levels of management will be explored. Participating agencies will be identifying opportunities for enhanced engagement with many stakeholders in the nanotechnology community—including regional, state, and local initiatives in nanotechnology and representatives from industrial sectors, nongovernmental organizations, and standards organizations at the national and international levels—as well as seeking mechanisms for targeted multilateral coordination on specific technical subjects. Moving into the next decade, meaningful engagement with stakeholders and ongoing external assessments will strengthen the efforts of the NNI as the participating agencies move toward realizing the four NNI goals.

Collaborative Agency Activities

As a multi-agency body, the NSET Subcommittee coordinates and collaborates on a variety of activities, such as nanotechnology signature initiatives that target multi-agency resources toward mutually agreed-on scientific and technological goals; development of joint research solicitations; and a wide variety of interagency meetings, workshops, and forums. In addition, member agencies work individually or in multi-agency collaborations in support of various R&D initiatives and dedicated facilities. The participation of member agencies in these activities varies with the relevance of any specific activity to the respective agencies' missions and goals.

Nanotechnology Signature Initiatives

To accelerate nanotechnology development in support of the President's priorities and innovation strategy, OSTP and the NNI member agencies have identified areas ripe for significant advances through close and targeted program-level interagency collaboration. This collaboration now includes nanotechnology signature initiatives that are intended to enable the rapid advancement of science and technology in the service of national economic, security, and environmental goals by focusing resources on critical challenges and R&D gaps. These activities also leverage skills, resources, and capabilities among various agencies in a concerted effort to maximize scientific and technological progress. The nanotechnology signature initiatives are being developed in the context of all four NNI goals. They are intended to genuinely affect the agency budget process, as encouraged by Administration guidance, and to dramatically improve ground-level functional coordination between agencies. The interagency groups supporting each initiative will identify thrust areas within each of the proposed initiative topics and identify specific agency programs that are involved. Finally, each nanotechnology signature initiative interagency group will select key research targets associated with near-and long-term expected outcomes, to help evaluate progress on an ongoing basis. The NSET Subcommittee anticipates participation and input from industry and other stakeholders on current and future nanotechnology signature initiatives. The first three nanotechnology signature initiatives[17] are described below.

Nanotechnology for Solar Energy Collection and Conversion

The President's agenda calls for the development of carbon-neutral alternative energy sources to mitigate global climate change, reduce dependence on foreign oil, improve the economy, and improve the environment. Long-term carbon reduction targets can and likely will be met by a portfolio of technologies, of which solar energy has the potential to play a prominent role. Solar energy is readily available, free from geopolitical tension, and not a threat to the environment through pollution or to the climate through greenhouse gas emission. The development of a solar energy infrastructure will not only support U.S. energy independence but also represents an unparalleled economic opportunity if the United States can maintain scientific and industrial leadership in this field. Today, the levelized cost of energy generated by solar technology is not yet economically competitive with conventional fossil fuel technologies without subsidies. Therefore, new innovations and fundamental breakthroughs can help accelerate the development of economical solar energy technologies that surpass the limits of existing technologies. Nanotechnology can help overcome current performance barriers and substantially improve the collection and conversion of solar energy. A number of nanoscale physical phenomena have been identified that can improve the collection and conversion of solar energy. Nanoparticles and nanostructures have been shown to enhance the absorption of light, increase the conversion of light to electricity, and provide better thermal storage and transport. However, current demonstrations of these technologies fall short of potential performance because of poor control over feature size and placement, unpredictable micro/nanostructure, poor interface formation, and in many cases, short lifetimes of laboratory devices. The goal of this initiative is to exploit the benefits of nanotechnology by enhancing understanding of energy conversion and storage phenomena at the nanoscale, improving nanoscale characterization of electronic properties, and helping enable economical nanomanufacturing.

Table 3. Agency Contributions by Thrust Area: Nanotechnology for Solar Energy Signature Initiative

Thrust Area	DOE	NIST	NSF	NIFA/ USDA	NASA	IC/ DNI
Improved photovoltaic solar electricity generation	•	•	•	•	•	•
Improved solar thermal energy generation and conversion	•	•	•	•		•
Improved solar-to-fuel conversions	•	•	•			

Sustainable Nanomanufacturing: Creating the Industries of the Future

The promise of establishing a significant number of new, high-value industries based on the past decade of investment in the NNI will be realized only if suitable manufacturing technologies can be developed to economically and reliably produce nanotechnology-based products on a commercial scale. The

---

[17] See http://nano.gov for the latest information on NNI nanotechnology signature initiatives.

semiconductor industry has achieved this, but the production methods are not scalable or economical for the diversity of new materials and products at the volumes and length scales required: radically new approaches are needed. Moreover, for such products to be ubiquitous in the nation's future economy, they and their associated manufacturing processes must be sustainable by design. To create the foundation for achieving this vision, the goal of this initiative is to accelerate the development of industrial-scale methods for manufacturing functional nanoscale systems. The initiative targets production-worthy scaling of three classes of sustainable materials (high-performance structural carbon-based nanomaterials, optical metamaterials, and cellulosic nanomaterials) that have the potential to affect multiple industry sectors with significant economic impact. The formation of consortia with industry, government, and academic representation is a key aspect of the specific material thrusts.

An essential prerequisite for the development of cost-effective nanomanufacturing is the availability of high-throughput, inline metrology to enable closed-loop process control and quality assurance. The initiative is therefore focused directly on the development of inexpensive, rapid, and accurate measurement techniques. The U.S. has expertise in roll-to-roll manufacturing, which can be adapted to the types of high-volume fabrication processes envisioned. The formation of a consortium devoted to the development of metrology methods to enable roll-to-roll application to nanomanufacturing is expected to play an essential role here. The systems to be manufactured based on these methods will include disruptive technologies for lightweight, high-strength, sustainable materials, solar energy harvesting, waste-heat management and recovery, and energy storage. Success of the initiative will result in the immediate extension of the methods developed to more complex components and systems as future nanodevices mature and will help secure and strengthen the U.S. manufacturing base.

Table 4. Agency Contributions by Thrust Area: Sustainable Nanomanufacturing Signature Initiative

Thrust Area	NIST	NSF	EPA	NIH	DOE	OSHA	NIOSH	FS/USDA	IC/DNI	NASA
Design of scalable and sustainable nanomaterials, components, devices, and processes	•	•	•	•	•	•	•	•	•	•
Nanomanufacturing measurement technologies	•	•		•		•	•		•	

Nanoelectronics for 2020 and Beyond

The semiconductor industry is a major driver of the modern U.S. economy and has accounted for a large portion of the productivity gains that have characterized the global economy since the 1990s. Recent advances in this area have been fueled by what is known as Moore's Law scaling, which has successfully predicted the exponential increase in the performance of computing devices for the last 40 years. This gain has been achieved due to ever-increasing miniaturization of semiconductor processing and memory devices (smaller and faster switches and transistors). Continuing to shrink the dimensions of electronic devices is important in order to further increase processor speed, reduce device switching energy, increase system functionality, and reduce manufacturing cost per bit. However, as the dimensions of critical elements of devices approach atomic size, quantum tunneling and other quantum effects degrade and ultimately prohibit the operations of conventional devices. Researchers are therefore pursuing more radical approaches to overcome these fundamental physics limitations. Candidate approaches include different types of logic using cellular automata or quantum entanglement and superposition; 3D spatial architectures; and information-carrying variables other than electron charge, such as photon polarization, electron spin, and position and states of atoms and molecules. Approaches based on nanoscale science, engineering, and technology are most promising for realizing these radical changes and are expected to change the very nature of electronics and the essence of how electronic devices are manufactured. Rapidly reinforcing domestic R&D successes in these arenas could establish a U.S. domestic manufacturing base that will dominate 21st-century electronics commerce. The goal of this initiative is to accelerate the discovery and use of novel nanoscale fabrication processes and innovative concepts to produce revolutionary materials, devices, systems, and architectures to advance the field of nanoelectronics.

Table 5. Agency Contributions by Thrust Area: Nanoelectronics for 2020 Signature Initiative

Thrust Area	NSF	DOD	NIST	DOE	NASA	IC/DNI
Exploring new or alternative state variables for computing	•	•	•	•		•
Merging nanophotonics with nanoelectronics	•	•	•	•		•
Exploring carbon-based nanoelectronics	•	•	•		•	•
Exploiting nanoscale processes and phenomena for quantum information science	•	•	•			•
National Nanoelectronics Research and Manufacturing Infrastructure	•	•	•			

Joint Research Calls

NSET member agencies develop joint research solicitations in areas of mutual interest to help address research gaps and leverage resources to achieve better research outcomes. The management of funding opportunity announcements (i.e., requests for application and/or proposals) is typically done by a lead agency, with participating agencies contributing research descriptions and names of technical peer reviewers. Following an external peer review, the individual agencies select and support separate research grants that are deemed meritorious and relevant to their respective missions and goals. Collaboration early in the process results in well-crafted solicitations that address critical scientific gaps in the identified area. Agencies are also able to leverage resources used to develop, publish, and manage research solicitations, thereby freeing resources for other needs. In some critical areas of mutual interest (e.g., nanotechnology-related environmental, health, and safety research), NSET member agencies have issued joint, bilateral research calls with research agencies of foreign governments. Future collaborative research solicitations will seek to build on these bilateral research calls and to engage a breadth of partners, including industrial and nongovernmental organizations and government research agencies in multiple nations.

Joint Research Facilities

NSET member agencies also develop joint research facilities that can combine strengths and expertise of participating agencies to enable faster and/or greater progress in nanotechnology research and development. One notable example to date is the Nanotechnology Characterization Laboratory, a collaboration of NCI, NIST, and FDA launched in 2005 to accelerate the transition of basic nanoscale particles and devices into clinical use by providing critical infrastructure and characterization services to nanomaterial developers. The development and support of collaborative research facilities going forward will be modeled upon such successful collaborations to enable broader participation from multiple agencies and industrial sectors.

Interagency Meetings, Workshops, and Fora

The NNI membership finds value in events that help bring together representatives from multiple agencies, as well as researchers supported by various agencies, to share knowledge and accelerate progress. NSET Subcommittee representatives help to disseminate information about topical meetings hosted by individual agencies, e.g., annual agency nanotechnology grantee meetings. Where multiple agencies participate in research solicitations, these grantee meetings feature information on the research progress from many agencies. An example is the EPA, NSF, NIEHS, NIOSH, and DOE Interagency Nano Grantees Workshop, held November 9–10, 2009, in Las Vegas, NV.[18] Moreover, agencies represented on the NSET Subcommittee may lead or co-sponsor studies to evaluate the current trends, opportunities, and gaps in nanoscale science and engineering R&D to aid policies and decisions about the NNI agencies' research investments.[19] Symposia and technical sessions organized by representatives from member agencies at various professional and technical conferences are also broadcast to the entire NSET community for participation. A number of NNI member agencies also participate in international member organizations, workshops, and fora, where they help to represent the United States and foster interactions with international partners.

Anticipated Activities

New activities planned for the immediate future include implementation of additional nanotechnology signature initiatives, leveraging collaborative interagency opportunities, working with the broad nano-

---

[18] For information on inter- and single-agency nanotechnology-related meetings and workshops see http://nano.gov.

[19] One example of such a study is *Nanotechnology Long-term Impacts and Research Directions: 2000–2020,* led by NSF and co-sponsored by other NNI participating agencies (see details at http://wtec.org/nano2/).

technology community to develop a robust hub for nanotechnology information dissemination, and an effort by the NNI leadership to strengthen support for the NNI throughout the Federal Government by engagement at all levels of agency management.

Building on lessons learned through the planning and implementation of the first three nanotechnology signature initiatives, future initiatives will be developed to address additional critical research needs and focus areas. The NNI will use nanotechnology signature initiatives as one means of addressing and meeting new challenges to our nation, society, and the environment using nanotechnology. The continued development of joint research solicitations is another way that the NNI will be able to incorporate emerging issues into its activities. These collaborative activities enable NSET member agencies to assess and meet new challenges at the outset.

Taking advantage of interagency collaboration mechanisms, many future activities will help mature the field of nanotechnology. For example, SBIR and STTR programs will benefit from joint agency funding opportunity announcements targeted in areas ripe for the maturation of nanotechnology-enabled concepts. Multi-agency collaboration in support of research facilities will also help advance nanoscale science and engineering. The NNI member agencies will continue to develop the applied science and expertise relevant to regulation needed to support responsible development of nanotechnology-enabled products. These efforts will be supported by the communication and collaboration engendered by the coordinated NNI framework of the NSET Subcommittee and its interagency working groups.

The NNI additionally recognizes the need for an internet-based, "one-stop shop" access point for nanotechnology information. Through this hub, various stakeholders and members of the nanotechnology community should be able to access a wide variety of information portals addressing nanoscale science and engineering education opportunities, nanotechnology-related careers, nanotechnology-based products, scientific data such as characterization and toxicity measurements, manufacturing instrumentation and resources, regulatory information, and other important elements of the nanotechnology enterprise. While some of this information is already found at http://nano.gov, the participating NNI agencies appreciate the need for a comprehensive hub and are committed to meeting this challenge. Initially, this will be done through individual efforts on the part of the agencies as their specific missions and mandates dictate. Achieving this long-term goal will require the assistance of stakeholders from industry, academia, nongovernmental organizations, and state and local governments, among others. International organizations and governments will also have a role to play in the establishment of a robust nanotechnology hub.

The NNI member agencies, with help from the NNCO, plan to strengthen support for the NNI throughout all levels of the Federal Government through a number of actions, including:

- Performing an ongoing mapping exercise to evaluate how this strategic plan relates to member agencies' strategic plans and to the priorities of the Administration.

- Holding meetings between the NNI leadership (i.e., NSET Subcommittee co-chairs, NNCO Director and Deputy Director, and/or working group co-chairs) and top-level management of each NNI member agency, to facilitate and strengthen agency support for the NNI, to discuss how the NNI activities can integrate better with R&D programs of each agency, and to become better informed about the goals and activities of each member agency with respect to nanotechnology.

Finally, NNI member agencies will continue to identify formal and informal mechanisms to overcome obstacles to interagency collaboration, which can arise due to differing agency needs, missions, cultures, and processes. Within these limitations, efforts will be encouraged to nurture relationships by drawing upon knowledge and expertise across agencies and by detailing agency staff to NNCO. These activities help to strengthen connections between agencies and support the NNI vision. Agencies are also individually encouraged to explore new forms of partnerships and collaborations.

Developing Partnerships and Engaging Stakeholders

As described in the Goals and Objectives section of this strategic plan, engagement with an array of stakeholders is considered to be critical to the future success of the NNI. The NNI will pursue effective methods to create and foster public private partnerships. Future plans for stakeholder engagement will involve the use of new media tools and interactive platforms such as the NNI Strategy Portal.[20] NNCO will play a more significant facilitating and coordinating role in interagency and agency-driven public engagement. A number of agencies already incorporate public engagement in their steering and advisory groups. Policy tools such as prizes and challenges sponsored by one or more Federal agencies can help spur nanotechnology innovation by engaging entrepreneurs, investors, universities, foundations, and nonprofit organizations. Furthermore, NNI member agencies will continue to participate in international standards organizations and multilateral fora to address policy-relevant nanotechnology issues and to promote international cooperation in aspects of nanotechnology that might affect human health and environmental safety.

Planned Independent Assessments

The NNI is regularly assessed by external advisory bodies, as shown in Figure 2. Recent reviews by PCAST and the NRC/NA have served to inform the strategic planning of the NNI as evidenced in this document and the *NNI Environmental, Health, and Safety Research Strategy*. The next NNAP review of the NNI is scheduled for 2012 under the current terms of the Act; the next NRC/NA assessment is scheduled for 2011.

Concluding Remarks

As indicated by the objectives outlined under each of the four NNI goals, the NSET Subcommittee developed this strategic plan as a means of moving toward achieving the NNI vision. By making concerted and coordinated efforts towards these goals through the stated objectives, the NNI agencies can together realize a vibrant nanotechnology R&D program. Continual internal assessment of progress is planned, in addition to the in-depth review and analysis of the NNI that the NSET Subcommittee conducts every three years during the revision of the strategic plan. The ultimate aim of this plan is to enable the NNI

---

[20] http://strategy.nano.gov is an online community that was originally established as a portal to solicit public input into this NNI Strategic Plan.

to move the country towards a strong, healthy, and prosperous future, capitalizing on the potential of nanotechnology as a stepping stone to attain this promise.

Over the next 10 years, we will continue to see new nanotechnology-enabled products, systems, and procedures with significant improvements in performance and functionality. Ongoing support for fundamental nanotechnology R&D and an awareness of responsible development will lead to new discoveries. The NNI represents not only a continued and dedicated investment of U.S. research funding for nanotechnology, but a thoughtful and encompassing set of processes for planning, coordinating, and evaluating such national investments. Moreover, the NNI's strong and sustained efforts to support education, public outreach, societal concerns, and stakeholder involvement will ensure that nanotechnology's potential is maximized and risks appropriately mitigated so that all Americans share the rewards of its benefit. Aside from the need for sustained support from NNI member agencies, the success of the NNI depends upon the insight and expertise of the broad stakeholder community (including, for example, academic researchers, industry representatives, and public citizens) as we continue to support research toward *a future in which the ability to understand and control matter at the nanoscale leads to a revolution in technology and industry that benefits society.*

Appendix A. External Assessment and Stakeholder Input

As referenced throughout this plan, a number of external independent sources provided the NNI with advice and recommendations during the creation of this document. In contrast to the many public NNI-sponsored workshops held in 2008–2010 on a variety of nanotechnology themes, the resources below targeted the development of this document. Finally, the results from a 30-day public comment period on the draft NNI Strategic Plan, November 1–30, 2010, are available for public inspection at the NNI Strategy Portal (http://strategy.nano.gov) and informed the final version of the NNI Strategic Plan.

External Assessment Reports

Report to the President and Congress on the Third Assessment of the National Nanotechnology Initiative is available at http://www.whitehouse.gov/sites/default/files/microsites/ostp/pcast-nni-report.pdf.

> The President's Council of Advisors on Science and Technology performed the third assessment of the NNI, releasing its report on March 12, 2010. The report is largely supportive of the NNI and contains a number of specific recommendations that have been incorporated into this plan, where possible.

Review of the Federal Strategy for Nanotechnology-Related Environmental, Health, and Safety Research is available at http://www.nap.edu.

> While the 2009 National Research Council's review document is mostly relevant to the NNI EHS Research Strategy, it did inform the drafting of some of the objectives in this strategic plan.

Stakeholder Input, July–August, 2010

To strengthen the development of this strategic plan, the NNI used a three-pronged approach to reach out to the nanotechnology stakeholder community for specific input. These activities occurred in July and August of 2010. The nanotechnology stakeholder community included those already familiar with the field of nanotechnology and the NNI, as well as those new to nanoscale science, engineering, and technology. Input was sought from a broad range of stakeholders, including members of the public; industry representatives; researchers in academic institutions; members of Federal, state, and local governments and regional initiatives; and representatives of nongovernmental organizations. The input from stakeholders in all three of these activities was invaluable in the development of the NNI Strategic Plan. Recommendations from the community have been carefully considered in creating the objectives found in this document. This plan does not make reference to specific R&D priorities beyond high-level areas such as those described in the nanotechnology signature initiatives section. Furthermore, each of the NNI member agencies separately determines its budget for nanotechnology R&D in support of its individual agency mission and needs. The NNI is an interagency budget crosscut in which agencies work closely with each other to create an integrated program through communication, collaboration, and coordination. The NSET Subcommittee will continue to use the stakeholder input regarding additional topics, such as R&D priorities and policy suggestions, to inform future decision making, as appropriate. The three stakeholder input activities were as follows:

- **Request for Information**
  A Request for Information (RFI) published in the Federal Register (Vol. 75, No. 128, Tuesday, July 6, 2010, pp. 38850–38853) referred to the NNI goals as a starting point for questions covering themes of Goals and Objectives; Research Priorities; Investment; Coordination and Partnerships; Evaluation; and Policy. Submissions were accepted from July 6–August 15, 2010. OSTP received responses from individuals, industry representatives, academics, state initiatives, and scientific societies.

- **NNI Strategy Portal:** http://strategy.nano.gov
  All stakeholders were invited to participate in the online public comment event hosted at the NNI Strategy Portal from July 13–August 15, 2010. Once registered, community members were encouraged to post original responses and to comment on postings by other members. In the online event, participants were invited to post responses in four timed stages in which the questions closely paralleled those posted in the RFI. At the closing of the online event on August 15, the NNI Strategy Portal community had almost 150 members. Although the period to respond to questions closed on August 15, 2010, the NNI Strategy Portal community continues to increase in membership, and it is seen as a potential mechanism for continuing to solicit public input in the future.

- **NNI's Strategic Planning Stakeholder Workshop**
  From July 13–14, 2010, the NSET Subcommittee held a public workshop in Arlington, VA, to solicit input from the broad stakeholder community regarding the development of the NNI strategic plan. The workshop included plenary lectures where subject matter experts shared their insights and discussed the status of nanotechnology research and application areas. The hard work of the workshop occurred during the breakout sessions, where participants were asked to help the NSET Subcommittee formulate specific objectives under each of the four NNI goals.

Some of the priorities, issues, and advice that the NNI received from stakeholders include the following:

- **R&D Priorities**
  Respondents identified a number of nanotechnology research priorities and concepts for future interagency nanotechnology signature initiatives, including quantum behavior, functional nanomaterials, photon-based computing, metamaterials, the nano-bio interface, tools for imaging and fabrication, process engineering, non-equilibrium systems, nanotechnology for low-cost sustainable energy, information needs, security, anticipatory governance, and nanotechnology pertaining to personal and public health (e.g., *in vitro* testing models, predictive toxicology, and high-throughput screening methods). The stakeholders were largely supportive of continuing and accelerating the NNI's efforts in nanomanufacturing, including the need for more research to increase the knowledge of fundamental processes and nanoscale phenomena.

- **Education and Workforce Development**
  Stakeholders emphasized the value of nanotechnology education as well as the critical need for a nanotechnology-trained workforce—including a recommendation to provide continuing education for patent examiners on the latest nanotechnologies.

- **Technology Transfer**
  Stakeholders provided examples of successful models for the transfer of technologies via state-led initiatives, gap funding, and other technology transfer mechanisms. The need to foster collaboration between industry, the Federal Government, and researchers at universities was articulated by many stakeholders, as was the need for the Federal Government's sustained support for the development of voluntary consensus-based international documentary standards.

- **Improved Interagency Collaboration**
  Strategies recommended by stakeholders to improve NNI interagency activities included grand challenge-themed pilot programs for interagency grant review boards involving program managers from various agencies, and grand challenge-themed strategy teams with shared personnel from multiple agencies.

For further details, readers are directed to the online strategy portal (http://strategy.nano.gov) and the report from the NNI Strategic Planning Stakeholder Workshop (available at http://nano.gov).

Public Comment on Draft Strategic Plan, November 2010

A draft of the NNI Strategic Plan was released for a thirty day public comment period on November 1, 2010. This was announced in the *Federal Register* (Vol. 75, No. 210, Monday, November 1, 2010, pp. 67149–67150 and at the online strategy portal (http://strategy.nano.gov). Responses and recommendations were received from the broad stakeholder community, including state public health, environmental, and regulatory agencies; entrepreneurs and industry representatives; academic researchers; scientific and occupational associations; and private citizens. All responses were made publicly available at the portal and were considered by the NSET Subcommittee in the final preparation of this document.

Appendix B. Glossary

the Act	The 21st Century Nanotechnology Research and Development Act of 2003
CDC	Centers for Disease Control and Prevention (DHHS)
CNST	Center for Nanoscale Science and Technology (NIST)
CoS	Committee on Science (NSTC)
CoT	Committee on Technology (NSTC)
CPSC	Consumer Product Safety Commission
DHS	Department of Homeland Security
DHHS	Department of Health and Human Services
DNI	Director of National Intelligence
DOC	Department of Commerce
DOD	Department of Defense
DOE	Department of Energy
DOEd	Department of Education
DOI	Department of the Interior
DOJ	Department of Justice
DOL	Department of Labor
DOS	Department of State
DOT	Department of Transportation
DOTreas	Department of the Treasury
EHS	environment(al), health, and safety
ELSI	ethical, legal, and societal implications (of nanotechnology)
EOP	Executive Office of the President
EPA	Environmental Protection Agency
FDA	Food and Drug Administration (DHHS)
FHWA	Federal Highway Administration (DOT)
FS	Forest Service (USDA)
GIN	Global Issues in Nanotechnology Working Group (NSET)
IC	Intelligence Community
IP	intellectual property
nanoEHS	nanotechnology-related environment(al), health, and safety
NA	National Academies

NASA	National Aeronautics and Space Administration
NCI	National Cancer Institute (DHHS/NIH)
NEHI	Nanotechnology Environmental and Health Implications Working Group (NSET)
NIEHS	National Institute of Environmental Health Sciences (DHHS/NIH)
NIFA	National Institute of Food and Agriculture (USDA)
NIH	National Institutes of Health (DHHS)
NILI	Nanomanufacturing, Industry Liaison, and Innovation Working Group (NSET)
NIOSH	National Institute for Occupational Safety and Health (DHHS/CDC)
NIST	National Institute of Standards and Technology (DOC)
NNAP	National Nanotechnology Advisory Panel (PCAST)
NNCO	National Nanotechnology Coordination Office
NNI	National Nanotechnology Initiative
NPEC	Nanotechnology Public Engagement and Communications Working Group (NSET)
NRC	National Research Council of the National Academies
NSET	Nanoscale Science, Engineering, and Technology Subcommittee of the NSTC Committee on Technology
NSF	National Science Foundation
NSRC	Nanoscale Science Research Centers (DOE program)
NSTC	National Science and Technology Council
OECD	Organisation for Economic Co-operation and Development
OMB	Office of Management and Budget (Executive Office of the President)
OSHA	Occupational Safety and Health Administration (DOL)
OSTP	Office of Science and Technology Policy (EOPH)
PCA	Program Component Area
PCAST	President's Council of Advisors on Science and Technology
R&D	research and development
RFI	Request for Information
SBIR	Small Business Innovation Research program
STTR	Small Business Technology Transfer research program
USPTO	U.S. Patent and Trademark Office (DOC)
USDA	U.S. Department of Agriculture
USGS	U.S. Geological Survey (DOI)

National Science and Technology Council
Committee on Technology
Subcommittee on Nanoscale Science, Engineering, and Technology